ALLE ZEIT WACH 1842

Dipl.-Ing. Dr. Vladimir Hubka
Institut für Konstruktion und Bauweise
Eidgenössische Technische Hochschule
HG-F-48
CH-8092 Zürich, Switzerland

W. Ernst Eder, MSc, P. Eng.
Department of Mechanical Engineering
Royal Military College of Canada
Kingston, Ontario, Canada K7K 5L0

This book is a completely revised English edition of
"Theorie Technischer Systeme", 2nd Edition, by Vladimir Hubka,
Springer-Verlag Berlin, Heidelberg 1984

ISBN 3-540-17451-6 Springer-Verlag Berlin Heidelberg New York
ISBN 0-387-17451-6 Springer-Verlag New York Berlin Heidelberg

Library of Congress Cataloging-in-Publication Data
Hubka, Vladimir. [Theorie technischer Systeme. English]
Theory of technical systems : A total concept theory for engineering design / Vladimir Hubka, W. Ernst Eder. p. cm. „Completely revised English edition of ‚Theorie technischer Systeme', 2nd edition ... 1984"-T. p. verso. Bibliography: p. Includes index.
ISBN 0-387-17451-6 (U.S.)
1. Engineering design. I. Eder, W. E. (Wolfgang Ernst) II. Title. TA174.H8513 1988

Printed in Germany

Dataconversion: Appl, Wemding
Offsetprinting: Buchdruckerei G. Buck, Berlin. Bookbinding: Lüderitz & Bauer, Berlin.
2161/3020-543210 Printed on acid-free paper

Preface

This book presents a comprehensive and unifying theory to promote the understanding of technical systems. Such a theory is useful as a foundation for a rational approach to the engineering design process, as a background to engineering education, and other applications.

The term "technical system" is used to represent all types of man-made artifacts, including technical products and processes. The technical system is therefore the subject (in the grammatical sense of the word) of the collection of activities which are performed by engineers within the processes of engineering design, including generating, retrieving, processing and transmitting of information about products. It is also the subject of various tasks in the production process, including work preparation and production planning, and in many economic considerations, company-internal and societal.

In this way, the Theory of Technical Systems is a contribution to science, as interpreted in the wider, Germanic sense of a "co-ordinated and codified body of knowledge". It brings together the various viewpoints of engineers, scientists, economists, ergonomists, managers, users, sociologists, etc., and shows where and how they influence the forms of engineering products. It also explains the influences that a product exerts on its environment.

This Theory of Technical Systems should thus interest design engineers, and engineers involved in production, management, sales, etc. In an interdisciplinary application of value analysis, the Theory of Technical Systems should provide answers to many questions raised in this field.

Keeping in mind the particular needs of students of engineering, we hope to reveal new insights to the relationships between different branches of knowledge for engineering with this book, and to encourage a better understanding of the range and nature of problems that engineers encounter.

This book pursues a number of aims:

- to describe and classify the principles of action of technical systems, and their properties and characteristics,

- to build up a basic terminology as a foundation for a study of engineering design, and
- to formulate important perceptions about technical systems on which to base further study of engineering design, particularly with respect to the design engineer's working methods.

This set of aims may look very dry and theoretical, but the reader's task should not necessarily be to study this material as a thorough and selfcontained treatise. On the contrary, after a first introduction, this book should be a constant companion, as a reference work for repeated search and discovery.

The first impulse towards compiling a theory of machine systems (as a special form of technical systems) came many years ago, from a study of design methodology. It was clear from the start that a generally applicable and well-founded design method cannot be built up without a clear model of an "abstract machine". Attempts at formulating such a model, and interacting with the developing design methodology, forced a continual revision of the ideas and perceptions. Many discussions with Dr. Hubka's colleagues in the Design Committee of the Czechoslovak Scientific-Technical Society helped to crystallize the ideas, particular help was received from Dipl.-Ing. M. Cervinka, J. Smilauer and S. Vit.

Further developments took place with friendly help and constructive criticism from Dr.-Ing. F. Kesselring. The result was the first edition of a book about theory of machine systems published in German: "Theorie der Maschinensysteme" by V. Hubka, Springer-Verlag, Berlin, Heidelberg, 1973 [109]. The second part of this discourse, a book about design theory, was also published in German: "Theorie der Konstruktionsprozesse" by V. Hubka, Springer-Verlag, Berlin, Heidelberg, 1976 [112]. In 1980, a further book appeared in German language: "Allgemeines Vorgehensmodell des Konstruierens" by V. Hubka, Fachpresse Goldach, Switzerland, 1980. It represented a development, the synthesis of the themes of design science, in a very condensed form. This volume has been translated into a large number of languages, including French, Italian, and Japanese. For the English language, a complete revision was undertaken, which among other things added a comprehensive glossary and index, and adapted the ideas more to the prevailing outlook of the English-language cultures: "Principles of Engineering Design" by V. Hubka (translated and edited by W. E. Eder), Butterworth Scientific, London, 1982 [119].

A full circle was almost reached, with publication in German of a second edition to the first of these books, under the revised title: "Theorie technischer Systeme" by V. Hubka, Springer-Verlag, Berlin-Heidelberg, 1984, sub-titled 'second completely revised and expanded edition'.

Since then, many discussions on these themes have taken place, particularly with Dr. M. M. Andreasen, and Dipl.-Ing. P. Ferreirinha. The main venue for such meetings has been at a series of conferences with the general title International Conference on Engineering Design: ICED 81 Rome, ICED 83 København, ICED 85 Hamburg, ICED 87 Boston. These conferences, and their planned sequels, are under the leadership of WDK-Workshop Design-Konstruktion, an informally constituted international society based on common interest in engineering design.

The Theory of Technical Systems has frequently been quoted in the world of scientific literature, which is a clear indication that the new discipline has proved useful.

The present volume is the result of active collaboration between both authors, and is again a complete revision and re-direction of its German pre-cursor, aimed specifically at the English-language audience. We have chosen to avoid any deep discussions about the works of other authors, even though their ideas have influenced this work, partly because of the limits of space set for the present book, and partly because it is aimed towards students of engineering. The problems should therefore not be clouded by added discussions. Similar practical reasons have caused us to select and document examples mainly from mechanical engineering.

The context for this book is set in chapter 1. Chapters 2 and 3 present the background on needs, and the transformations that serve to satisfy those needs. Chapter 4 contains information on processes and their properties. Technical systems and their properties are the subject of chapters 5-11. Chapter 12 summarizes some perceived applications for the theory of technical systems. All statements collected to summarize the chapters, and all summarizing propositions stated within the chapters are collected for easy reference in appendix A. Appendix B contains a discussion of some of the terminology, including the original chapter 2 from the pre-cursor of this book, and a discussion about choice of words, to make some of the philosophy clearer.

Thanks are due to the following for valuable comments to drafts of this book: Ken Wallace, Lecturer in Engineering Design, University of Cambridge; Geza Kardos, Professor, Carleton University; Charles O. Smith, Professor (retired), Rose-Hulman Institute of Technology.

The authors take full responsibility for errors or omissions and would like to hear any constructive suggestions and comments.

July 1988

V. Hubka, Greifensee, Switzerland
W. E. Eder, Kingston, Ontario, Canada

Contents

1 Introduction 1
1.1 Technology 1
1.2 Technology As Cultural Factor 3
1.3 Development of Interpretations on Technical Systems and Theories 4
1.4 Technical Systems 7
1.5 Theory of Technical Systems 8
1.6 Summary 10

2 Needs - Demands
Technical Requirements - The Job to Be Done 13
2.1 Case Study - Logging 13
2.2 Human Needs 19
2.3 Design to Fulfill Human Needs 22

3 Transformation Systems 23
3.1 Model of the Transformation System 23
3.2 Elements of the Transformation System 30
3.3 Transformation Systems - Statements 33

4 Technical Processes 35
4.1 Model of Technical Processes - The "TP Model" 35
4.2 TP Operands 41
4.3 TP Structure 42
4.4 Received Effects, Produced Effects, Effects 45
4.5 TP Secondary Inputs and Outputs 48
4.6 TP Operators 48
4.7 TP Characteristics and Evaluation 49
4.8 Systematics of Technical Processes 53
4.9 Operating Activities Connected with the Process 55

4.10 TP Representation 55
4.11 Special Theories of Technical Processes 56
4.12 Technical Process – Statements 57

5 Technical Systems 58
5.1 Nature of Technical Systems (Habitude) 59
5.2 Model of Technical Systems – "TS Model" 64
5.3 TS Function Structure 72
5.4 TS Organ Structure 77
5.5 Component Structure of Technical Systems – TS Component Structure 81
5.6 Comparison of TS Structures and Their Transformations 85
5.7 TS Boundary 90
5.8 TS Environment 90
5.9 Technical Systems – Statements 91

6 Classification of Technical Systems (Systematics) 93
6.1 Classification of Technical Systems by Function (Effect) 95
6.2 Classification of Technical Systems by Action Principles 96
6.3 Classification of Technical Systems by Degree of Complexity 97
6.4 Classification of Technical Systems from Manufacturing Similarity . 99
6.5 Classification of Technical Systems by Difficulty of Designing . . . 100
6.6 Classification of Machine Parts and Groups (Sub-Assemblies) by Production Location and Degree of Standardization 101
6.7 Classification of Technical Systems by Design Originality (Degree of Novelty) 103
6.8 Classification of Technical Systems by Type of Production 104
6.9 Classification of Technical Systems by Degree of Abstraction 105
6.10 Classification of Technical Systems by Type of Operand 105
6.11 Classification of Technical Systems by Application in the Technical Process 107
6.12 Classification of Technical Systems by "Quality" 107
6.13 TS Systematics – Statements 107

7 The Properties of Technical Systems 108
7.1 Characteristics of Technical Systems As a Class 111
7.2 Categories of Properties of Technical Systems 112
7.3 Relationships Between the Properties 143
7.4 Determining the Properties and Their Relationships 146
7.5 Specification of Properties of Technical Systems 151
7.6 Mental Processes of Realizing the Properties of Technical Systems . 156
7.7 TS Properties – Statements 158

8 Evaluation of Technical Systems 160
8.1 Evaluation Procedures 160
8.2 TS Evaluation – Statements 165

9 Representation of Technical Systems 168
9.1 Types of Representation for Technical Systems 168
9.2 Representation of Constructional Elements 172
9.3 Representation of Technical Systems - Statements 173

10 Origination and Operation Phases of a Technical Systems
(Ontogenesis and Genetics of TS) 174
10.1 Origination Phases of Technical Systems Manufactured in Quantity 178
10.2 Origination Phases of a Technical System Manufactured in One-Off Production 183
10.3 Origination Phases of a Technical System of Fourth Degree of Complexity (Plant) 185
10.4 Origination and Operation of Technical Systems - Time Dependency 186
10.5 Execution of Partial Processes 190
10.6 Origination of Technical Systems - Statements 191

11 Developments in Technical Systems in the Course of Time
(Phylogenics, Evolution of Technical Systems) 192
11.1 Regularities in the Long-term Development Process 192
11.2 Trends in Developments 201
11.3 Regulating and Controlling the Process of Long-term Developments 206
11.4 Motivation for Long-term Developments 208
11.5 TS Developments in Time - Statements 210

12 Applications of the Theory of Technical Systems 211
12.1 Special Theories of Technical Systems 211
12.2 Engineering Design 214
12.3 Further Applications of the Theory of Technical Systems 226
12.4 Relationships of TTS to Other Knowledge 231

Appendix A. Summary of Statements and Propositions from Chapters 233
Appendix B. Terminology 241
Bibliography 262
Subject Index 269

Letter and Graphical Symbols

a	abstract
AEnv	active environment
Aes	aesthetic property
Aux	auxiliary
B	behavior
c	concrete
C	condition
CPM	critical path method
D	newly designed
Des	design property
Di	distribution property
DP	delivery and planning property
E	energy
Ec	economic property
Ef	effect (exerted by TS)
Eff	effector
El	element, elementary
ElEf	elementary effect
ElTS	elementary technical system
Env	environment
Erg	ergonomic property
EV	effectiveness
FD	functionally determined property
Fu	function Fu_i function number i Fu^i function with i conditions
FuTS	function possessed by a TS
f	number of steps of complexity of TS in state of finality
g	weight factor
Hu	human (individual or group)
HuS	human system
i	serial count (number)
I	information
In	input
IS	information system
LC	law conformance property
Liq	liquidation property
Mc	cross-product of two-place relationship
M	material
MEl	machine element
Mfg	manufacturing property
M&GS	management and goal-setting system
MS	machine system (special case of TS)
n	number
Od	operand Od_i operand number i Od^i operand in state i
Op	operation
Opp	operational property
Ot	operator
Out	output
Pa	partial
PaEf	partial effect
PaOp	partial operation
PaTP	partial process
PaTS	partial technical system
Pb	problem
Pc	principle
PERT	program evaluation review technique
Pr	property Pr_i property number i Pr^i property in state i
Pro	production property
PuD	purchased (bought out, out-sourced), newly designed
PuSt	purchased, standard
PuTyp	purchased, typified or works standardized by supplier
PV	profitability
p	probability

R relationship
R/C regulation and control
Re re-used
Rec receptor
s relative strength
S system
Sec secondary
SIn secondary input
Sol solution
SOut secondary output
St standardized
T technical
Tg technology
TgPc technological principle
TP technical process
TrP transformation process
TrS transformation system
TS technical system
TSFu technical system that fulfills a function
TTS theory of technical systems
Typ typified
W working
WPc working principle
WSt works standardized
WTP working process

Graphical Symbols

 process or process system

 function

 decision process

 object system

 input or output

 effect

1 Introduction

1.1 Technology

Language deals with meanings, but the words used do not always have a single definition, or even a precise one. This is the case with the word *technology*, the title of this section, which is used in two different senses in this book. The German word *Technik* is more expressive and encompasses a wider range of meaning in a better definition. (Note: some further discussions and definitions of terms and expressions used in this book may be found in appendix B.)

Technology, in the wider sense of *Technik*, means "practical or industrial arts; ethnological study of development of such arts; application of science" [3], or similarly "a technical method of achieving a practical purpose" [9]. All manner of artifacts, how they work, how they are produced, the knowledge (science, and know-how) that they incorporate, the effects they have on society, etc., are therefore covered by this definition. Later chapters will consistently employ a much narrower definition of the word "technology" [115, 119], and avoid this wider one.

People who wishes to gain a view of technology, its range of activity and influence, and its place in and value to society, without starting from definitions and statistics, have to observe indirectly. They can try to recall relevant experiences from their own memory, by describing their activities during a "normal" day, and listing the items and procedures that concern any artificial (non-natural) products. If all the technical objects encountered in that time period are exhaustively listed – not only the most obvious such as shaver, car, train, street-car (tram), television, but also the less noticeable ones like toothbrush, spoon, knife, water tap, pot, stove, house, street, bridge, door lock and key, lighting fixture, chair and other furniture, writing implements, and many others – an impressive collection of items will be observed that show our dependence on technology.

Still more revealing is a visit to a technical museum. By walking several kilometers (or miles) around the exhibits, a good idea can be obtained about what technology offers, and into which areas of society it has penetrated. This process may even produce (as a side effect) some admiration for those persons who created these items of technology.

If you wish to elevate yourself through art, then the poem written by Rudyard Kipling (1865–1936), the Nobel Prize winner of 1907, will serve this aim:

The Secret of the Machines (Modern Machinery)

We were taken from the ore-bed and the mine,
  We were melted in the furnace and the pit –
We were cast and wrought and hammered to design,
  We were cut and filed and tooled and gauged to fit.
Some water, coal, and oil is all we ask,
  A thousandth of an inch to give us play:
And now, if you will set us to our task,
  We will serve you four and twenty hours a day!

    We can pull and haul and push and lift and drive,
    We can print and plough and weave and heat and light,
    We can run and race and swim and fly and dive,
    We can see and hear and count and read and write!

Would you call a friend from half across the world?
  If you'll let us have his name and town and state,
You shall see and hear your crackling question hurled
  Across the arch of heaven while you wait.
Has he answered? Does he need you at his side?
  You can start this very evening if you choose,
And take the Western Ocean in the stride
  Of seventy thousand horses and some screws!

    The boat-express is waiting your command!
    You will find the *Mauretania* at the quay,
    Till the captain turns the lever 'neath his hand,
    And the monstrous nine-deck city goes to sea.

Do you wish to make the mountains bare their head
  And lay their new-cut forests at your feet?
Do you want to turn a river in its bed,
  Or plant a barren wilderness with wheat?
Shall we pipe aloft and bring you water down
  From the never-failing cisterns of the snows,
To work the mills and tramways in your town,
  And irrigate your orchard as it flows?

    It is easy! Give us dynamite and drills!
    Watch the iron-shouldered rocks lie down and quake,
    As the thirsty desert-level floods and fills,
    And the valley we have dammed becomes a lake.

But remember, please, the Law by which we live,
We are not built to comprehend a lie,
We can neither love nor pity nor forgive.
If you make a slip in handling us you die!
We are greater than the Peoples or the Kings -
Be humble, as you crawl beneath our rods! -
Our touch can alter all created things,
We are everything on earth - except The Gods!

Though our smoke may hide the heavens from your eyes,
It will vanish and the stars will shine again,
Because, for all our power and weight and size,
We are nothing more than children of your brain!

(Reproduced from "A Choice of Kipling's Verse", T. S. Eliot (ed), London: Faber 1941, p. 294)

Admiration for the poet should be tempered by a further thought on the role of technology in society: once again it is technology, in this case the technical processes of producing books (paper-making, printing, cutting, binding, etc.), that enables wide distribution of something that is aesthetically pleasing, this poem in this book. Thus technology also supports culture and the arts.

1.2 Technology As Cultural Factor

Every human society is characterized by its *culture.* According to R. Thurnwald, this is the totality of habits, contexts and establishments that relate to family, social and political organization, economy, work, morality, tradition, justice, philosophy, etc. Cultural items can be grouped into the following broad areas:

- **economic**, including sub-groupings of agriculture, mining, animal husbandry, *technology*, industry, trade and transportation;
- **social**, including tradition, justice, society, state and education;
- **intellectual**, including sciences, arts, religion, literature, language and games.

Culture works through the media of technology. A group of people who are collectively equipped with abilities and knowledge to increase their stock of technology and knowledge are said to have a *civilization.*

The entire life of human beings is closely tied to the accomplishments of technology. The self-evidence with which the human being uses the technical achievements is a special testimony for this close relationship between man and technology. Technology has made its impression on the whole of our nature, and no doubt has consequently had deep influences on our mental constitution, our cultural life, and finally on our total culture.

1.3 Development of Interpretations on Technical Systems and Theories

The human being has the mental ability to disregard unimportant or accidental characteristics, properties and relationships, and to emphasize the significant and important ones in order to focus on the observed object or process in a deeper and truer fashion. This is one of the most valuable characteristics of human, and it has helped humanity to attain its leading position in nature. Abstracting (as a mental process) has helped *homo sapiens* in all areas to obtain knowledge, to group and classify it and provide it with suitable structure, and thus to create and expand individual sciences, and even to create meta-sciences.

When a problem had to be mastered, there was not always a theory available. Technology is full of instances where practice has preceded theory, and the subsequent development of theory has served to improve the practice. This is true for the Theory of Technical Systems, presented in this volume in its current state of development.

Let us pursue the development of *the perceptions about a machine.* In earlier times, each machine was viewed as a totality that consisted of parts that belonged and were peculiar to that machine. A water mill was called the "upper mill" or the "lower mill" according to its location, but with almost no regard for elements that were common between the two. Older books therefore described individual machines as a whole, with almost no reference to similarities in other machines. There was even a deficiency of usable words for some machines, for instance Ramelli in 1588 lacked a generic name for "pump".

The processes of discriminating only began with the founding of polytechnical schools, the earliest of which were Paris 1794, and Prague 1806. Mechanisms were the first elements to be separated from the study of the entire machines (Monge, Carnot, Hâchette and Lanz), by defining at first ten, and later 21 classes of mechanisms for transforming motions in the machine. Borgnis (1818) went further by distinguishing six classes of machine-organs (in a loose analogy with terminology from the biological sciences) that were ordered according to their functions (what they would do), rather than according to motion transformations. These thoughts about functions and organs were accepted and developed by Coriolis and Poncelet, who recognized three essential parts to every machine: receptor, transmitter, and effector (tool). For instance, a screw driver can easily be visualized in this way, the handle is the receptor, the tool shank is the transmitter, and the blade end is the effector. This interpretation was rejected by the representatives of a rapidly disseminating study of machines (Ampère took part in this movement), and was forgotten.

This sketch of early developments would be incomplete without considering Leonardo da Vinci (1452–1519). He had already recognized the mechanisms and machine elements as consistently repeated elements of every machine, and studied them as such. In his comments (Madrid Codex I and II) he formulated two interesting postulates about machines:

- a book about the nature of machines must be written before a book about their application,
- mechanisms are a paradise for the mathematical sciences, mathematics is stimulated by mechanisms.

Leonardo da Vinci is recognized as a genius, he was one of the few people capable of anticipating the normal development of a science.

The systematic study of machine elements and mechanisms was only undertaken in the nineteenth century. By the start of that century a sufficiently large number of machines of reliable construction had been built, and it was then possible to search for the underlying regularities, patterns, themes and common elements. Apart from machines for warfare, mining, and water usage, some other machines existed such as spinning and weaving machines, machine tools, pressure and lifting machinery, and were soon followed by steam engines, electric motors and generators, gas- and liquid-fuelled internal combustion engines, etc. The designers and users of many of these groups of machinery used different terms to designate similar items, and developed their own terminology (see appendix B for some relevant dictionary definitions and discussions).

It is worth noting at this point that different languages followed varying trends in developing a terminology. Most languages acknowledge that to think about a new device is a matter of some ingenuity, and consequently that the person doing such thinking (a designer) is an *ingenium*, or at least an Ingenieur (as used in the German). The English language has relegated this connecting link between thought and ingenuity to designate merely the physical artifact that results from mental work; the embodiment of the ingenious idea is labelled an "engine", and by extension anyone who works with the (completed) machine is an engine tender, or engineer. It would be interesting, but beyond the scope of this book, to follow the cultural and sociological developments that caused this aberration (compare [59, 76, 77]).

The most pressing problems in the nineteenth century existed when engineers tried to establish adequate dimensions for parts. These problems were connected with the limited knowledge available in the general science of mechanics, and particularly with the mechanics and strength of materials. F. Reuleaux (1829–1905) saw mechanics, and especially kinematics, as the basis for finding the principles according to which machines work or function. It was Reuleaux who, based on the research of F. Redtenbach (1809–1869), attempted to formulate the first general theory of machines in his "Theoretical Kinematics: Basis for a Theory of the Nature of Machines". This interpretation has survived to this date in the "theory of machines", as exemplified in the works of P. Willes and P. L. Chebychev (compare Artobolevskii [33, 34]). Works in this spirit have also been presented in the U.S.A. [192]. They quote three areas in the theory of machines and mechanisms: synthesis of mechanisms, machine dynamics, and theory of automata.

Such a limiting view of the theory of machines and mechanisms was not generally known in Europe, but the interpretation described above is acknowledged (even if only by tradition, and sub-consciously) in teaching and in the structure of the sciences. For instance in mechanical engineering, the traditional scientific structure of curricula, and their elements (courses) and the relationships between them, still have as their basis the disciplines of mechanics (including strength, fluids, thermodynamics, etc.), machine elements, materials and manufacturing technology, and techniques of representation (mainly graphical). A theory to connect these areas did not exist at that time.

Individual areas certainly developed alongside this general stream, particularly with respect to the basic sciences. Examples of sciences that developed in this way are: thermodynamics applied to thermal machinery, fluid dynamics applied to fluid machinery, and textile technology for textile machinery. For practical reasons, all questions concerning planning, manufacture and use of equipment had to be solved almost independently within the framework of each area of expertise that produced such a product line. The result was a series of closed, self-sufficient bodies of knowledge in which many years of experience were required to obtain the technical education in each professional or occupational area. The main reason was the absence of a general (or meta-) theory, resulting in the unsystematic collection of "know-how" without the structure that such a theory can provide. A contributing factor to the scattered collection of knowledge was the existence of separate organizations within engineering and between countries, which made transfer of knowledge and experience between specialities more difficult.

Such a situation was only acceptable in the early phases of the first industrial revolution, in which the transition from manual craft-based manufacture to industry-based production (and the age of machinery) took place. With continuous increasing industrialization, various crisis situations arose, in particular during the second World War. These, and the problems of ecology and raw material shortages, and problems arising from the trends of automation in the second technical revolution demanded new departures and theories. During this second technical revolution new technical tools and "high technology" were created, and enormously increased demands were placed on equipment. Coupled with the availability of new aids to finding solutions (e.g. computers), these circumstances have suggested a need for a new scientific basis of knowledge about technical systems.

It is somewhat surprising how little systematic knowledge exists about the nature of technical objects such as tools, appliances, machines, etc. This becomes particularly obvious when technology is compared to other areas of science which investigate and categorize items such as minerals, animals and plants into complicated systems, taxonomies and sub-systems.

A new approach to technology emerged from the stress of the second World War, namely to attempt to formulate a more general theory to describe all technical systems, and to recognize "the machine" in its context as a system. At first, this approach only existed as a treatment of a few aspects of a general theory within the framework of other thematically related writings [133, 210]. It appeared later in a more integrated form, particularly connected with cybernetics and systems thinking, for instance:

1962 Gosling: **The Design of Engineering Systems** [90]

1968 Hubka: **Theorie der Maschinensysteme** (prepared for press in Czechoslovakia in 1968, issued in German in 1973) [109]

1970 Roth: "Systematik der Maschinen und ihrer Elemente", section concerning functions [186]

1974 Hansen: **Konstruktionswissenschaft**, chapter on the theory of technical objects [100]

since 1974 Yoshikawa: various works [202, 212, 213]

1976 Ropohl: "Ansätze zu einer allgemeinen Systematik technischer Systeme", [184]

Since that time the theory of technical systems presented in this book has been developed, through a second edition of Hubka's work and many other efforts. This theory has been acknowledged as a necessary basis, and as a source of information, for unifying and integrating the knowledge of various technical areas. In particular, design science rests on the theory of technical systems [64, 64, 67, 69, 110, 119, 126, 127].

1.4 Technical Systems

When visiting a technical museum, we can see thousands of technical objects and we recognize them as products of technology. Their variety in functions, forms, sizes, etc. tends to obscure the common features and properties among these objects. But a vast variety also exists in nature, and in those circumstances clearly defined kingdoms of natural objects were formulated as subjects for investigation in the various natural sciences. Let us therefore, as a first step in presenting the theory of technical systems, attempt to develop a term that conceptually describes all classes of technical objects.

Technical objects can be referred to as objects, products, things, machines, implements, technical works (Dessauer), etc. The results of a manufacturing activity, as the conceptual content of technology, can be termed artifacts (Kapp), or *instrumentum* (Heidegger). These definitions are intended to include all manner of machines, appliances, implements, constructions, weapons, ships, etc. that represent the "technical means" (Heidegger) by which the human being achieves his "ends".

500 years ago it was conceptually difficult to combine paintings, musical creations, literature, poetry, sculptures and so on under the collective term "works of art". Similarly today it is difficult to find a suitable term in technology for all technical works. It is particularly difficult because we are dealing not only with a wide variety of form, function and complexity, but also with a multitude of principles on which they are based. Much of the candidate vocabulary has already acquired a "specific narrow" meaning, having been used for some time in various disciplines, though the content and meaning is usually only determined by feel, and not by formal definition.

There are two possible ways to find a suitable expression for the concept of an "abstract machine" and other "technical means". Either one can attach a comprehensive meaning to the word "machine", and thereby attempt to include all types of machine products, e.g. plant for sugar refining, washing machine, gear box, thermometer, transistor, shaft, screw. Alternatively one can search for a new expression. Here we have chosen the second alternative, and all "technical means" are designated as *systems* with particular attributes. These terms have already been accepted, even though they are not uniformly used, and thus any further arguments are unnecessary. The following sections serve to define these terms more precisely.

Starting from the term "system", a hierarchical sub-division of systems can be achieved according to figure 1.1. Only one aspect of the origin of systems is considered here, differentiating in the first step between artificial systems and natural ones.

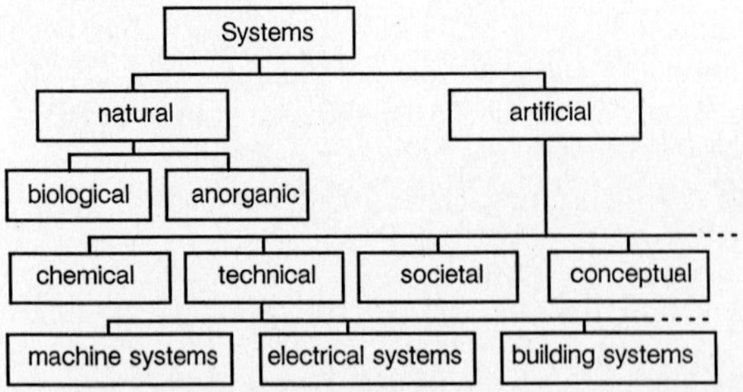

Fig. 1.1 Hierarchy of Systems according to their Origin

With the help of this hierarchy we can see the boundaries of the individual terms much more clearly, because they deal with established areas of knowledge. A problem still exists within the classification of technical systems. The classes shown in this scheme coincide with the established sub-divisions of technology and industry in mechanical engineering, electrical engineering, civil engineering, etc. This classification from a practical and organizational viewpoint does not permit a precise definition of a machine system or electrical system, because no firm boundary can be drawn by describing these systems as products of mechanical or electrical engineering. Modern developments of technical systems have generally tended to blur the boundaries, electrical and computer products are now increasingly used together with mechanical and civil engineering products, each acting as sub-systems to more complicated transformation systems.

Even if we classify systems according to the appropriate principle of action, as mechanical, electrical, hydraulic, pneumatic, etc., we obtain no uniform criteria, nor do we obtain advantageous classes of technical systems, because most modern systems are hybrids.

These reasons have led to the circumstance that the term *technical system* is being used more and more for designating the "abstract machine", and is therefore adopted for this book, frequently abbreviated to "TS".

1.5 Theory of Technical Systems

The "theory of machines" is an established area of knowledge, as was explained in section 1.3. Technical systems (TS) are not just mechanisms, and they are not just automata. Consequently the "theory of machines" as interpreted by Artobolevskii [34] can not provide complete coverage of the field, because it deals with only a part of the total theory.

Let us then introduce the concepts of a new interpretation, a comprehensive Theory of Technical Systems (TTS), which in its current state of development forms the subject matter for this book.

1.5.1 Aims of the Theory

The primary aim of the Theory of Technical Systems presented in this book is to classify and categorize the knowledge about technical systems (as the subject of the theory) into an ordered set of statements about their nature, regularities of conformation, origination, development, and various empirical TS-related observations. Simultaneously, a suitable terminology (a system consisting of terms as its elements, and their definitions which describe the relationships among terms and to general language) should be created such that their meanings do not need further explanation, and that additional terms can be derived from the basic group. The form of deductive science to which we aspire can only be achieved in a number of discrete steps. This volume represents the work to date, as an entry point to further developments.

1.5.2 Structure of the Theory

The structure of the theory of technical systems is derived from the important questions that the theory should answer through its statements, and which are the subject of most of this book. These questions are about:

- terminology, a system of terms, and their applicability to all technical systems
- transformation systems
- technical processes (TP) – as major elements of transformation systems
- technical system (TS) – object systems as elements of transformation systems
- goal systems – system of purposes (goals, aims) for the TS
- conformation of the TS, elements (of various kinds) and their relationships
- properties of the TS, evaluation of the TS with respect to its properties
- origination of the TS (ontogenesis) – from idea or need, through full manufacturing specification and usage to disposal
- development of a TS during passage of time (phylogenesis)
- systematics, types and families of TS

1.5.3 Types of Theories

Theories of technical systems can be distinguished according to their range of validity:

- a general theory that is valid for all TS and machine systems, the theory presented in this book
- many special theories which provide more concrete statements derived from the general theory with respect to individual phyla, classes and families of TS.

The divisions proposed here for the special theories follows from a loose analogy with biological taxonomy, dividing the general theory into a suitable number of levels. The special theories can form a hierarchy within the general theory, thus yielding a theory of machine tools, a theory of lathes, etc. A particular place can be reserved for those theories that are applicable across the interface between a number of areas, such as the theory of mechanisms, machine elements, electrical elements, and others.

We use here an analogy with the subdivisions and hierarchical levels established in biological taxonomy, and have adopted some of the relevant terms. Further explanations will be presented in section 6.9.

1.5.4 Relationship of the Theory to Other Disciplines

The theory of technical systems should build on the existing set of sciences, and relate them in an ordered fashion. It should also help to develop new sciences whose number is increasing with the range of usages for which technical systems are being applied, and with the enhanced demands made for using TS. Apart from the "classical" sciences such as physics (with all its constituents) and chemistry, engineering draws more and more on biology, system theory, economic sciences, ergonomics, logic, cybernetics, and many others. The theory of technical systems presented in this book in turn should form a framework and classifying system (a meta-theory) for many engineering sciences that are connected with designing, manufacturing, testing, selling, storing, transporting, operating and disposing of TS. These more specific engineering sciences provide added detail to the statements of the general theory of TS. For example:

- "mechanics of materials" (strength) is the science that investigates the relationships between resistance to load (as one of the properties of a technical system) and the geometric and material properties of the TS; similar duties are performed by the knowledge of "reliability", "life-span", "manufacturing technology", etc.: they each deal with a limited range of questions, a *scientific detail* area of the general theory of technical systems.
- "fluid dynamics" describes and researches the processes of fluid flow: it is regarded as a *special process theory*, one of many within the interpretation of the theory of technical systems.
- "theory of mechanisms" treats mechanisms as partial systems, or sub-systems as parts of a TS: it is consequently regarded as a *special theory of technical systems*, a constituent part of the general theory of technical systems.

1.5.5 Application of the Theory of Technical Systems

The theory of technical systems is intended to have wide applicability to technical practice, and should also have a considerable significance in cognitive theory. The treatment of a system of object-oriented disciplines, such as the individual (specialized) areas of technology, carries many consequent effects - it should provide a unified order, and clear relationships and boundaries between individual disciplines. Based on these properties of the theory of technical systems, it should have particular importance in general education in the regions of technology and society, and especially in engineering education where this theory of technical systems can and should serve as the integrating factor.

1.6 Summary

This chapter can be summarized by stating a few reasons for formulating the theory of technical systems as presented in this volume. Advantages are also gained by a close link with established systems theory and cybernetic theory.

- The theory of technical systems delivers relationships that are claimed to be valid for all technical means (including products). It should serve to assist in transferring technological experiences from one area to another, based on the relationships between categories of systems and homomorphism that exist between technical objects.

- Classifying all technical products uniformly as technical systems should enable us to develop working methods for engineers that are independent of product, and transferable between areas of endeavor. We can thus learn and teach the processes and contexts of designing technical systems, and not only the design of pumps or lifting equipment.
- Working with abstract terms should tend to force engineers to use scientifically founded methods, because the traditional imaginative powers of designers are by themselves no longer sufficient to create better products in this high-technology age of complex systems. This creates a prerequisite for the ability to overcome obsolete traditional methods and patterns.
- Systems thinking incorporated into the theory of technical systems presents the opportunity to treat problems as a whole. This is a necessary pre-condition for consistently successful design and other engineering effort.
- The connection of the theory of technical systems to cybernetics and its terminological framework provides engineers with better communications to other scientists. It eases the tasks of formalizing many logical operations within the design process, especially as increased use of computers during engineering design depends on formulating algorithms for those design operations where such logical treatment is possible. Therefore progress in computer-aided design techniques should be closely connected with the theory of technical systems. Introducing CAD in an *ad hoc* way, driven by the existence of CAD hardware and software rather than by the needs of the designer, can no longer meet the needs of advanced engineering. The needs of the designer, and the science of engineering design, are an extension of the theory of technical systems.

  Note that the theory of technical systems, and its extension in the theory of design processes (design science), also acknowledges those human operations that are not strictly logical in this context, and thus are not predictable or algorithmizable. The latter include the intuitive and creative leaps that frequently occur, which are responsible for many technical advances. The results of these design operations cannot be relied on for immediate correctness, and must like all other results be checked and verified. Nevertheless, intuition and creative leaps belong to the design engineer's working methods.
- Formation of classes of technical systems based on similar behavior offers engineers a basis for finding a larger number of technical means that will fulfill a particular behavioral requirement or function. This produces the conditions which permit finding the best solution to a problem from the available variety of possible solutions. A practical form of information for this purpose consists of the design catalogs, formulated to collect and present technical and scientific information in patterns that are more useful for engineering design, e.g. [187], as tables of principles, operations, and available solutions.

The theory of technical systems is introduced, and progressively extended, in the following chapters. The nature and roles of needs and their fulfillment are outlined in chapter 2. Chapter 3 discusses the socio-technical transformation system, and the place of engineering and design in the social context. Chapter 4 defines technical processes, describes their characteristics, and shows the nature of their operators. The operator that is most amenable to control by individuals and groups in society, the technical system, is the subject of chapters 5–11. Especially those features that are common to all technical systems are highlighted. The discussion in chapter 5 deals with the nature of technical systems, their constituents, and structures and models that can be set up to describe them. Chapter 6 shows various ways of classifying technical systems by many of their features. The properties of technical systems are featured in chapter 7. Chapter 8 outlines how technical systems can be evaluated, both from the viewpoint of engineering in the quantitative technical-scientific aspects, and from the viewpoint of the user and society in the qualitative aspects of attitudes and impact on lifestyles. Some discussions of formal representation of technical systems in models (including drawings) is presented in chapter 9. Chapters 10 and 11 show how technical systems behave during time, one aspect treats their development from idea and need to a full manufacturable description of a system, the other aspect concerns their historic evolution with the state of the art. Chapter 12 contains some envisaged applications of the theory of technical systems, and shows the current stage of development of the science about engineering design. The brief outline presented in section 1.5 above is expanded as far as possible within the limits allowed by this book.

2 Needs - Demands

Technical Requirements - The Job to Be Done

Technology (in the wider sense) can be regarded as the sum of all those real and potential processes occurring in the world which are influenced by human beings. They include acquiring and extracting raw materials and foodstuffs, manufacturing and transporting products, building houses and bridges, producing information media, and a vast number of others. Also included are many service tasks such as banking, retail sales, personal services, etc., and the equipment used for them, machines, tools, devices, appliances, computers, and so on. Let us consider one process from this almost infinite variety as an example. It will serve briefly to illustrate a few basics of the theory of technical systems, and provide a foundation for understanding the theoretical analysis which follows in subsequent chapters.

2.1 Case Study - Logging

Wood is one of the most useful and widely available materials that humans have procured from nature since time immemorial. The techniques and processes connected with harvesting and using wood and trees are probably among the most widely known. Most people have had some personal experience of them. Who, for instance, has not gathered or cut wood, and prepared it to make a camp fire?

2.1.1 Description of "Logging" from Encyclopedia

Let us start with an established way of obtaining information, by reading appropriate entries in an encyclopedia [1, 2, 4, 5, 7], for instance under a heading such as "logging", or "forestry". A typical short entry reads as follows [1]:

Logging
Logging is the practice of harvesting and transporting logs from forest lands to wood-manufacturing facilities. The size and nature of the equipment used in log-

nology of logging as practiced today. It shows the contributing technologies and the technical means used in them, and deals with some of the effects and individual functions, i.e. "what work does it do?" "what does the work?" "how?" and "with what purpose?" This way of defining the technology determines to a large extent the technical requirements (specifications) that must be met by the technical means. We can show this by a few examples. By considering some alternative sequences of operations, the possibilities of alternative solutions (and technical means) can be highlighted.

(d) Sequence of Operations
The description in the quoted article prescribes a viable sequence of the transformations. It is nevertheless possible that the operation of trimming can be performed as first or as third operation. Such a new technology demands quite different technical means. If the standing tree is to be trimmed of its branches, the resulting machine to perform this task has to be capable of vertical positioning, for instance by climbing up the trunk (see Figure 2.3). If alternatively the trunk is to

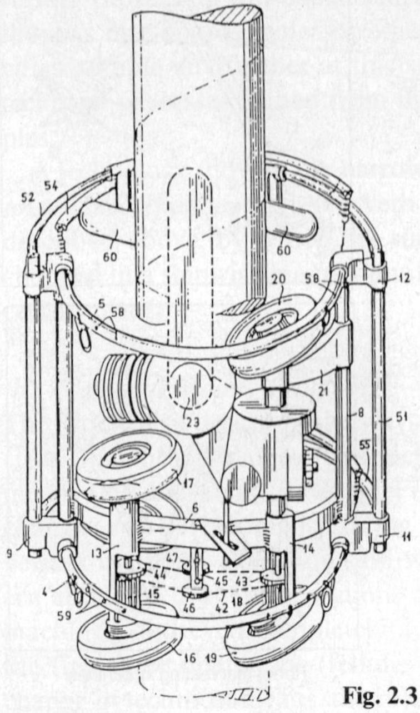

Fig. 2.3 Climbing Saw Machine (Drawing from Patent 444565)

be bucked into logs before the branches have been removed, or even transported in this state, a different set of equipment will be needed. These latter alternatives may at first sight seem to be absurd. Yet equally absurd would be to select a conventional solution, and not to examine and consider such alternative technologies. In fact, both types of machines have been developed about 20 years ago by a Canadian company, and used in suitable locations.

(e) Mode of Action

If we speak about the *technology* of cutting, then the *mode of action* of the technical means used for separating the tree from the ground (how the technical means work) is already largely established, namely severing the wood fibers by sawing or by shearing. A very similar result can be achieved by breaking the trunk, or by pulling the trunk and roots out of the ground. By selecting such alternative principles of "separating the tree from the ground", a whole set of different technical means are naturally applicable, other than the axe, or the hand- or motorpowered saw that was originally implied in "cutting and felling trees".

(f) Description of Process

A further point can be made with this example, namely about how to describe a technical system. Language and verbal statements are not generally suitable to provide an easily assimilated description of complex processes. The contents of the verbal description can be simply transformed into a graphical representation, figure 2.4. The power of diagrams to describe processes or products becomes obvious, without at this stage discussing any details or principles of representation and modelling. As Confucius said, "a picture is worth a thousand words".

a) Diagram of Logging Process

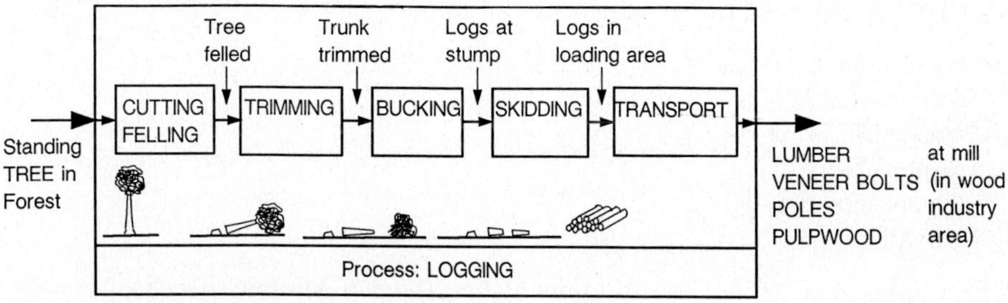

b) Historical Developments of Technical Means

Historical Stage	Technical Systems	Power	Control
1	Axe, saw, wedge, cable	Human, draft animal	Human
2	Chain saw, tractor, hoist truck	Motor	Human
3	Automated tree feller, balloon, helicopter	Motor	Human + auto-controller

Fig. 2.4 Diagram of Transformation Process "Logging", and Historic Developments of Technical Means

2.2 Human Needs

All persons perceive some deficiencies in the world around them, which are usually coupled with a wish to alleviate the perceived problems. These perceptions are labelled *human needs*, and satisfying these needs can be declared as the goals of that person or society.

Various kinds of needs exist. Sociologists have discussed the nature of human needs, and the priorities that people place on them. For instance Maslow [153, 154] recognizes at least five levels of need. These may be summarized as follows:

(1) **physiological**, such as food, clothing, shelter from cold or rain, etc., existential needs that are connected with survival and maintaining life;
(2) **safety or security**, including physical and emotional aspects;
(3) **self-respect**, love, comfort;
(4) **esteem**, respect from others, prestige, status;
(5) **self-expression**, self-actualization.

The physiological need is obviously the most basic of these needs. Such needs form a hierarchy, leading in steps to higher levels of satisfaction. Once a lower need is sufficiently satisfied, a higher need can receive attention. If any circumstances arise that threaten a lower need, the human being reverts to trying to satisfy that one, and will usually neglect the higher one. The limits are relative, what one person may regard as sufficient satisfaction may be entirely inadequate for another.

Hertzberg [103, 104, 105] identified a different dimension of human behavior concerning needs. Various factors act as *motivators* for people to satisfy certain needs, they are listed (in reducing order of importance) as:

- achievement,
- recognition,
- work itself,
- responsibility,
- advancement, and
- growth.

They seem to be mainly related to the highest three of Maslow's levels of satisfaction. Other factors must be present as *hygiene factors.* They do not motivate if they are present in sufficient quantity and quality, but de-motivate if they are lacking. These factors include:

- company policy and administration,
- supervision,
- relationship with supervisor,
- work conditions,
- salary,
- relationship with peers,
- personal life,
- relationship with subordinates,
- status, and
- security.

The two lower levels of Maslow's hierarchy of needs figure in this list. This extends the discussion about how adequately a need must be satisfied, and what affects a person's performance in satisfying needs. A detailed discussion is beyond the scope of this book, more may be found in [47, 51, 55, 103, 104, 105, 153, 154, 197, 211].

Generally each change in development of a society or social group brings about a different set or assortment of needs. In the middle ages, people saw no need to travel from (say) London to New York in five hours, or to own a weekend cottage on a nearby lake, or to have a parking space available near the home or workplace. In contrast, today one finds no need for a sword or a horse (except as a recreational implement at the higher levels of self-fulfillment). As our situation changes, some needs remain, others are dropped, others change their character, and some new ones emerge.

Needs are given various levels of priority that depend on one's perception of the immediate or possible future situation. Everyone has a system of needs (or a system of goals) that consists of individual needs and the relationships between them.

The relationships between any two needs can be neutral if fulfillment of one need does not affect the nature of the other. Alternatively, the relationship can be competitive or conflicting, if one need affects another in a negative way, or synergistic if one need affects the other positively. The relationship can act equally in both directions, or it can act differently in each direction. A *causal relationship* exists between a need and the means to fulfill that need. For example, the need for a washing machine has a causal relationship to the need for clean clothing. See appendix B for a discussion of different types of relationships.

These considerations suggest an important experience, namely that each need can be satisfied by various different means, and in a number of different manners. The need for clean clothes can be satisfied not only by washing them in a machine, but by many other activities, including buying new clothes.

Needs, and the selection of means to fulfill them, is guided by the value system of the society. What is recognized, valued, honored or aspired to by the dominant culture is elevated to the preferred needs or goals of life.

In general, people will tend to formulate their needs in terms of the existing technologies, available devices, and peer-group attitudes. The introduction of a novel device on the market can radically change the perception of our needs, a process known as *device, process or technology push.* This phenomenon is aided by advertising – a social and informational process that attempts to stimulate needs by inducing people to think about a product and regard it as desirable. Alternatively, a particular need expressed by a person or group can induce someone to try to satisfy it – a process known as *market pull.*

A large number of additional factors and considerations influence humans and their expressed needs. These include fashions and fads, attempts to keep up with the perceived state of the art, peer group influences (attempts to keep up with the neighbors), political, economic and ecological concerns (e.g. the energy crisis, or zero population growth), etc. Within a broad social structure, there are generally diverse patterns of needs between social sub-groups. Differences can be found depending on the value system of the smaller sub-groupings. They are neither constant in time, nor necessarily cohesive within an identifiable group of people.

Humans find very few completely adequate means in nature to satisfy their needs. The available natural means must be altered, changed, increased, transformed, transported, processed, in a multitude of ways, so that the individual can obtain adequate means to satisfy the perceived need. We must cook, bake, add salt, melt, cut, dismantle, transport, store, generate energy, sow, fertilize, reap, grind, stack, measure, calculate, and much more. By all these processes, an *operand* (the material, energy and information that goes through the process) is given additional values, properties and qualities, such that the final *state* of the operand satisfies the previously declared need. The end state may be reached through a number of intermediate states. We will use the term *transformations* to describe these processes of change.

The means actually chosen by people to fulfill their needs depend on the technologies available at that time, the costs of the technical means, and the benefits that the persons expect from them. A product must thus be judged from a number of viewpoints, related to marketability, quality of design and production, simplicity, cost of acquisition and operation, safety, etc. Many of these factors can not be measured by an individual on a specific device - hearsay and reputation play a role in one's perceptions about the suitability of a product.

2.3 Design to Fulfill Human Needs

If the means to fulfill a need do not exist at the time, or only exist in an inadequate form, but are capable of being realized, then a process of designing and manufacturing a new product can supply the means to fulfill that need. This would constitute an example of market pull as mentioned in the previous section. The first stages in providing these new means are usually labelled **product planning** (a management task) and **designing** (an engineering task), which can as occur separate or overlapping processes, and usually need iterative repetition.

Engineering and technology are not only "subjects" for study, they are also the means by which new and improved products (technical systems) are devised and made. The Theory of Technical Systems should therefore also be applicable to the processes that generate the information needed to make these products, the design process. Further discussions on these topics are contained in chapter 12.

Design engineers aim to achieve the best possible results in the design process, particularly by achieving optimal quality of designed products (TP and TS) in the shortest time at minimum design costs, especially with respect to low life-cycle costs, human and societal acceptability, etc. This is the engineering design problem.

In presenting the theory of technical systems in this book, we aim not only to describe the features common to all technical systems, but also to point out various aspects of engineering design. They consist of advice and suggested procedures that can help the designer towards producing better technical systems. The resulting outline of engineering design is by necessity incomplete, a fuller discussion is available in [29, 119, 121, 122].

3 Transformation Systems

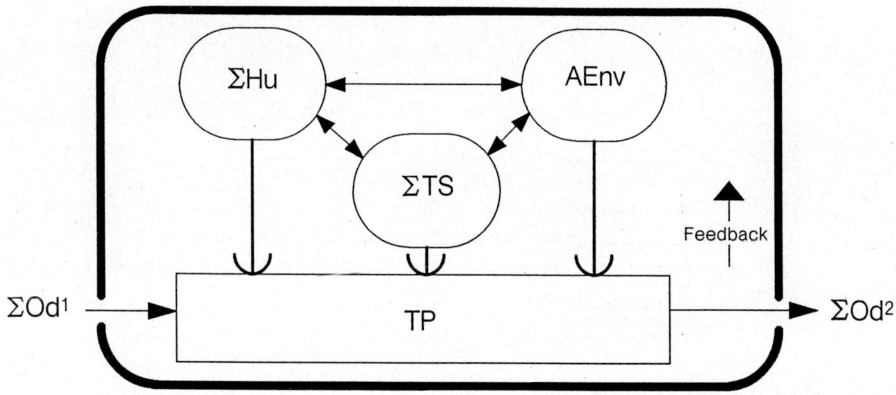

3.1 Model of the Transformation System

In chapter 2 we described a certain class of processes under the collective term "transformation", and defined some terms and characteristics. At this stage, a number of further terms and symbols are introduced, and briefly brought into context. More precise and complete definitions and discussions of some terms may be found in appendix B, and the symbols are listed in the preface section of this book. The remainder of this book will serve to give fuller explanations of these concepts, and progressively expand the scope to show the theory of technical systems. It will also show some of the consequences of this theory for engineering design and designers, although this aspect is covered more thoroughly elsewhere [112, 119].

The particular example of "logging" quoted and briefly discussed in Chapter 2 may be graphically described by figure 3.1a. The general model shown in figure 3.1b can be derived by abstracting from this example (and many others). This model is fundamental in the theory of technical systems described in this book. It incorporates elements of a systems-theory to explain the role of technology in society proposed by Ropohl [185], which shows that parts of society are themselves modellable as socio-technical systems.

The sum of all elements and influences (and the relationships among them and to their environment) that participate in a transformation is collectively termed a *transformation system.* A clear overview of the general way in which these elements can be classified and ordered can be obtained from figure 3.2.

Each transformation system has a fairly well-defined *purpose,* namely to perform the intended transformations on the appropriate operand, and thus fulfill the stated and implied needs. One of the purposes of engineering design (within the socio-technical system) is to provide the information about suitable real processes, and their operators, which should be capable of realizing the transformations to fulfill the needs, namely by changing a suitable operand from its starting to its

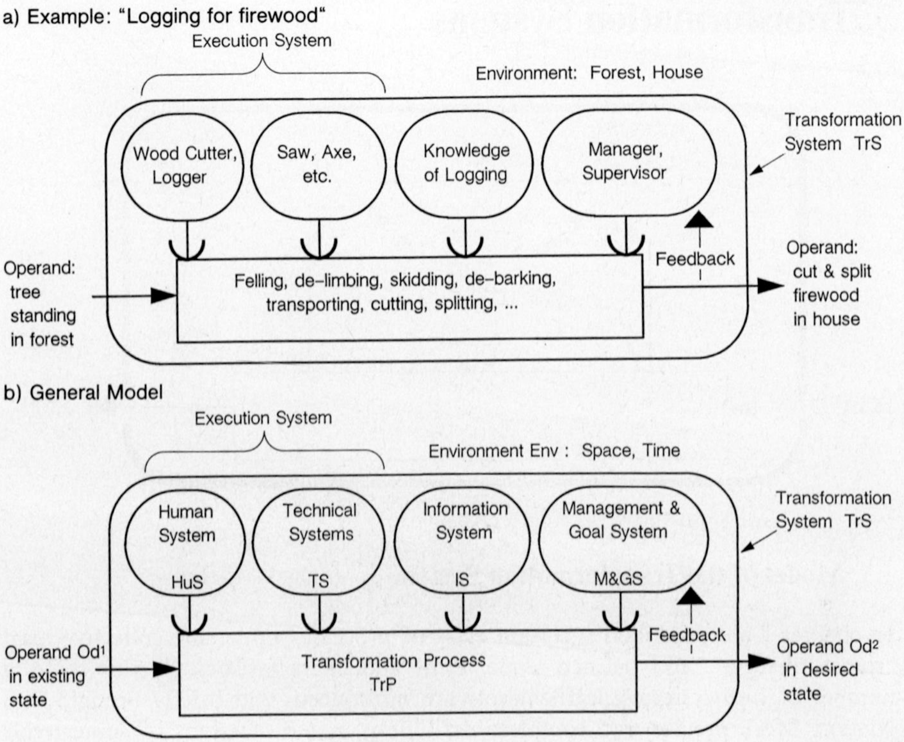

Fig. 3.1 Model of the Transformation System – Specific and Abstract

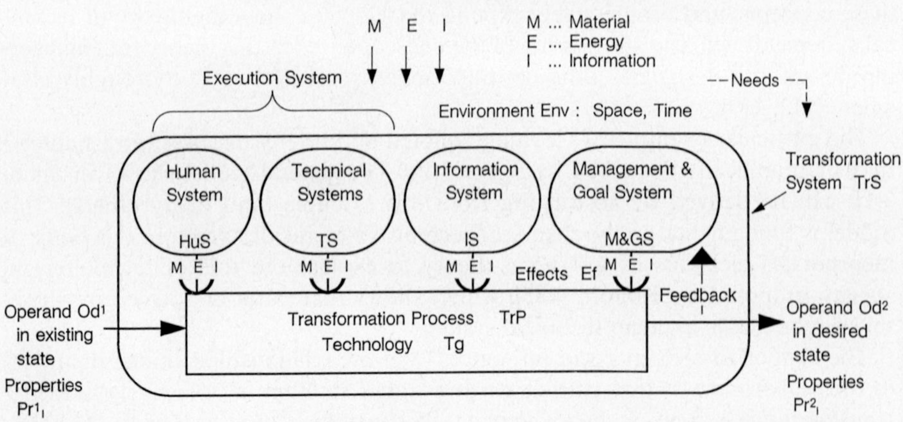

Fig. 3.2 Model of the Transformation System – Elements and Examples

final state. The purpose of the transformation system described in the example of logging in chapter 2 was to change the operand "tree" from its starting state "standing in a forest" to its final state "lengths of tree-trunk near a sawmill (plus various by-products)".

Separable *elements* of a transformation system (defined in appropriate ways) are

	Transformation General $Od^1 \rightarrow Od^2$	Technology Tg	Neccessary Effects Ef (and Examples)	Contribution of Operators to Effects (and Examples) HuS	TS	AEnv
Examples	1 Steel object soft → hard	Hardening	Heating (in Furnace)	Worker (temperature measurement)	Furnace	Air ← Heat →
			Quenching (in water)	Transport of component	Water tank	Heat
	2 Grain grow one → many	Grain seeding in prepared ground (action–medium), harvesting, & storing	Ground loosening and turning	Driver (steering)	Tractor Plough	Ground
			Seed planting metering	Driver	Seed Drill	
			Weed destroying	Worker	Cultivator	
			Germinating			Sun
			Growing			Rain
			Reaping	Driver	Harvester	
			Threshing			
			Storing		Silo	
			Transporting	Driver	Truck	
Examples	3 Material raw material → private car	Designing	Concept finding	Designer	Computer	
			Drawing	Designer	Drafting machine	
		Work preparing	Choosing technology	Technologist		
		Manufacturing	Metal cutting	Worker	Mach. tool	Workshop
		Assembling	Part connect	Assembler	Press	
		Testing	Measuring	Tester	Measuring equipment	
			Adjusting	Mechanic		
Examples	4 Human London → Paris	Flying	Forming a runway	Pilot	Wing profile	Runway
			Lift		Fuselage	
			Protecting humans			
		Road travel	Forming a road			Road
			Moving		Wheels	
			Propelling		I.C. motor	Fuel pump
			Steering	Driver Police	Steering system	Traffic control

Fig. 3.2 (cont.)

connected by suitable *relationships* within the system. These will be discussed in more detail in a subsequent chapter of this book (see also appendix B). The example of logging showed various elements (machines, personnel, etc.) and their relationships (operations, operational sequences, management directives, etc.).

The properties of the whole system result from the sum of the elements, but the system's *behavior* includes the synergistic effects of the relationships. The connections and arrangements of the elements and relationships form a set of recognizable *structures* for the system. Some of the structures for the logging example show (a) the operations and their sequence that accomplish the transformation of the operand, (b) the machines available for assisting the humans to perform these operations, (c) the management and control arrangements that guide the transformation system.

The totality of measures (numerical values) for the *properties* of a system or an individual element at a certain time is termed its *state.* Connections exist from the system to its *environment* by means of the system's *inputs* and *outputs.* The terms used in this paragraph need further explanations, but that is one of the purposes of this book.

Referring to figure 3.2, the major elements of the total transformation system are conveniently divided into a *process*, an *operand* that is being transformed, and the *operators* that drive and guide the process. An analogy to the concept of a mathematical function and its operator, operand and result is intentional.

The technical *process* has *inputs* and *outputs*, both of which contain desired and unwanted elements. Its *characteristics* include the purpose of carrying out a transformation of an operand. The *operand* can consist of *materials*, both biological (including the human being) and nonbiological, *energy* and *information* (commands and data). The total transformation may involve many operands of all types, the word "operand" used in the singular is intended to include all operands. The transformation from the entry state of the operand (at the input to the process) to its exit state (at the output) takes place during a period of time in a certain space. The transformation may affect the operand by altering its *structure* (internal characteristics), its *form* (external characteristics), its *location* (space, e.g by transportation) and/or its *time* (e.g. by storage). Space and time are not normally considered as operands, but as general properties. Time, in general, can not be arbitrarily influenced, it elapses at its own pace. Space is a property of both form and location.

The process has an internal *structure* that consists of separable elements, termed partial transformations or operations. These are connected by relationships, such as compatibility of output from one operation with input to a following one, time, etc.

The *operators* consist of human beings, technical systems, information systems, management systems and the immediate environment. They individually exert *effects* to drive and guide the process. The exerted effects may be of material, energetic and/or informational (command or data) nature. In the logging example, the human operators are the loggers, their helpers, their supervisors, and higher management. The technical systems include chain saws, winches, wedges, notebooks, information files, computers, etc. The effects they exert include applying the teeth of the saw chain to the tree-trunk, laying and tying ropes to logs, etc. The management system decides which trees are to be cut, and exerts its effects mainly as instructions (command and data information). The environment includes rain, and government regulations about logging and re-foresting.

The way in which a particular effect interacts with a particular part of the transformation is termed a *technology* (now in the narrower sense compared to that used in Chapter 1). Any one effect can be achieved (or realized) by a variety of technologies. Each technology (including information technology) is based on one or more natural (including sociological) phenomena that provide the principle for operation of that technology, the *technological principle.* These principles are mainly described by the various sciences.

The duties of *setting goals* and *achieving goals* are normally performed by the management system. Such goals are not usually individual statements of what is (or is desirable) to be achieved. More usually they are an interconnected set of statements, a *goal system* that is set by the management system (acting as a *goal-setting system*), and achieved or realized by the actions of all operators in the transformation process.

Implied in this model are superimposed processes of *feedback* which involve measuring the output of a (total or partial) system or process, comparing this to a desired standard or goal, and dynamically altering the input to attempt to correct any detected error or deficiency. The dynamic behavior of the processes may be stable, or problems of instability may occur. Such feedback takes place for the operand, for the process, and for each of the operators – it is thus the duty of internal and external feedback loops for each system. Feedback is one of the relationships between elements of a transformation system.

A *technical system* is usually an artifact, a product of human art and workmanship. The effects that it exerts onto the operand of a technical process are termed the output functions of that technical system (what it should and can do when it is made to work). They include by implication any external and internal feedback functions to help to control the system.

The required properties of a technical system must be attained (by design, manufacture, development, etc.), and the technical system consequently carries those properties that should serve industry and mankind.

The *environment* can be divided into that part that immediately affects the transformation system, the active (close) environment, and the remainder that has little or no effect on the system, the passive (remote) environment.

The distinction between a "process" and "technical system" may seem artificial, especially because a separable process on a distinct operand may not readily be visible. Nevertheless, such a distinction has been found useful especially for engineering design.

Such a division into process and operators is not unique but depends on the viewpoint of the observer. A designer may wish to look at a thermal power station as an energy conversion system (with energy as the main operand) to generate electricity from coal, and using a water pumping system as one of its operators. Alternatively, attention may focus on causing a forced flow of water (the operand), and increasing its pressure as the process of interest, and the pump impeller and housing, electric motor, switch, wiring, and so on constitute the operators. In another view, the designer may wish to look at the connection between the shaft and the impeller, and designate transmission of energy (the operand) from one to the other as the process of immediate interest. Further examples of transformations, technologies, effects and the activities of operators are also listed in figure 3.2.

Any system is always a constituent part of a super-system, and can itself be divided into sub-systems; systems constitute a *hierarchy.* Each part of the scheme shown in figure 3.2 can in turn be similarly modelled as a system.

A part comprising a single piece of material ready for assembly into a larger technical system can be regarded as a system within that hierarchy. It has inputs and outputs (typically force and motion), internal characteristics (for instance non-uniform stress, inertia), properties (strength, mass, appearance), elements (of form or geometry, of crystals and atoms), and relationships between these elements. It is a part of a larger system, and can be divided into smaller parts (but these are generally no longer of interest to the mechanical engineer).

This model of a transformation system, which is based on experience, is thus built on the following premises:

- The necessary transformation of the operands is achieved by applying specific (purposeful) effects which may be classified into material, energy and information types.
- These three types of effect are delivered to any transformation from the following sources: a set of humans, a set of technical systems, a set of information systems, a set of management and goal systems, and the active environment.

The whole system of the world, with its human societies and other relevant elements of the solar system, permits us to consider an almost infinite set of transformation systems. Individual elements frequently participate in a number of different transformations at the same time, for instance a person may be sitting in the restaurant compartment of a railway train (process of transportation), eating, reading a newspaper, and simultaneously discussing a political situation with a companion.

Society (the whole or its parts) may be regarded as a system in this sense (compare [54, 184, 185]). Its elements are individual human beings, groups of humans recognized by common occupation or purpose, and artifacts and organizations within and around which these humans act. The relationships include culture, economics and financing, politics, etc. Socio-technical transformation systems can therefore include manufacturers, suppliers, distributors, planners, foremen, users, reporters for communications media, spectators, innocent victims, etc., as well as those systems that are primarily technical, such as cars, machine tools, printing presses, distribution equipment, etc.

A socio-technical system generally should be of some practical use, it should perform a duty by driving and guiding a process to fulfill a human need. By performing its duty the system changes a part of its environment, and will thereby also change a part of itself. All transformation systems are thus basically *dynamic.*

Those parts of the socio-technical system that represent mainly human actions and interactions (e.g. goal-setting and goal-realizing) cannot be fully described by the input/output nature of this model. An additional model for autonomous human behavior and organizational action and behavior (termed "closure" in the sociological and management literature) can explain some of the differences [88]. Human organizations tend to develop loyalties and self-motivation within groups, and tend to make decisions that are not deterministic or apparently rational. Self-interest and other actions of the individuals also plays a part, especially in decision-making.

Examples of such socio-technical systems may be found in any company and its management, any bank (an important transformation in this system is that of money from one account to another), local governing councils, and so on. Such systems are studied in detail within the subjects of management, sociology, political science, and others. It should be relatively easy to find the elements and some of the more obvious relationships within these spheres to fit the model of transformation system as described here.

The model of a transformation system shown in figure 3.2 illustrates the earlier statements about transformations. A short summary is as follows (see also statements 3.1 to 3.11 at the end of this chapter):

- In order to satisfy the direct and indirect needs of an individual or group of people, an object (an operand - Od) that exists in a suitable state (a desired state Od^2) is selected, or in some way pre-determined by circumstances. The exit state (Od^2) is the goal to be achieved by the transformation.
- Four possible classes of objects are available as operands of the transformation: biological objects (including in particular humans), material, energy, and information.
- The design engineer searches for a suitable state of the operand as the input to the transformation (or it may be determined by circumstances). The existing state of the operand (Od^1) is an unsatisfactory one. In most cases a number of starting states (Od^1) are available from which the exit state (Od^2) can be attained.
- The changes made on the operand to convert it from the entry state (Od^1) to the exit state (Od^2) is the transformation.
- The transformation uses (is based on and driven by) a technology (in the narrower sense) that depends on some natural phenomena, and defines the partial transformations, and the operand's intermediate states.
- In most cases, a number of technologies are available to perform the planned transformation $Od^1 \rightarrow Od^2$.
- The transformation or partial transformation is realized (made into a reality, produced) by the effects (material, energetic, informational) acting on the operand (compare Statement 3.1 at the end of this chapter).
- The effects on the operands are exerted by the operators: humans, technical systems, information systems, management systems, and the active environment.
- All the operators are coupled to one another and to the environment by material, energetic and informational links.

This model serves generally for all *types of transformations* on all kinds of operands and with any operators producing the effects that act (by means of the technology) on the operands, as demonstrated by the examples in figure 3.2.

If the "internal workings" of one of the operators are analyzed, various processes can be found that themselves could be considered as transformations. Typical transformations *within* a technical system (one of the operators of a transformation process) are: converting rotational motion into linear motion, amplifying a force, etc. The aim for a technical system is to produce at its output a particular effect on the operand. The purpose and character of this change within the technical system is qualitatively different from the transformations performed by the technical system on the operands as outlined in an earlier paragraph of this section. It is therefore preferable if the internal transformations within a technical system (or other operator of a process) are termed *action processes*, processes that result from an internal action. The connected elements of that TS are termed an *action chain*, a chain of elements that is capable of performing an action when it is suitably activated.

A *causal relationship* [172] thus exists between the transformation process (performed on the operand of a process) and the action process (taking place within an operator). Namely the changes of an operand ($Od^1 \rightarrow Od^2$) in the transformation process (TrP) are caused by the effects exerted by the technical system. These causes (the output effects of the TS) are in turn a consequence of the action chain within the TS, the causes of which are the inputs to the TS. Further explanations to this theme will be presented in chapters 4 and 5.

It is generally easier to recognize the causes from observed consequences, a process of diagnosis, than to predict consequences from existing or envisaged causes, a process of planning and design. In turn, physical phenomena are more easily predictable (under the usual limiting and simplifying assumptions that can lead to mathematical formulations of theories) than general, social and human phenomena. Human choices and peergroup interactions are among the least predictable phenomena, partly because they involve "free choice" by individuals [173]. Nevertheless, the idea of causal relationships is very powerful for human purposes.

3.2 Elements of the Transformation System

A transformation system consists of the elements shown in figure 3.2. Some additional comments with respect to the individual elements are useful.

The *operand* is a passive member of the transformation system. It can belong to one or more of the following classes: biological material, non-biological material, energy and information. In almost all cases these occur in combination, and it is almost impossible to separate them, but one or other may be regarded as the principal operand, and the others can usually be neglected. The state of the operand is characterized by the totality of measures of all its properties, but can also be characterized by the values of separate parts of the operand, especially if the relevant transformation is that of "combining" or "dividing". The transformation process as the connecting link in the transformation system is discussed more extensively in chapter 4, under the title of *technical process.*

The *human as operator* is that sub-set of all humans who directly exert any effects on the operand of that particular transformation. The *technical system as operator* is that sub-set of technical systems that directly exert any effect on the operand of that particular transformation. These two are the main elements of the *execution system* that causes the effects to drive the process.

Details of technical systems are extensively discussed in chapter 5. They are the key to many problems, and can be surprisingly extensive for some relatively simple transformations. Example 4 in figure 3.2 names transportation as a process of "changing location", and even the simple part of that journey, the ground-transport, needs many technical systems including a car, roads, bridges, fuelling stations, garages, maintenance facilities, and many more. It is only when these operators exist that this process of transportation can definitely and reliably take place.

The *management and goal system* acts indirectly to drive the transformation process. It constitutes the *executive system*, with the purpose of providing coordinated direction of the execution system to achieve a desired end. The *information system* is used by both the execution and the executive system as storage medium and source for necessary information.

The *active environment* as operator comprises all sources of effects exerted by the surroundings, most of which are not explicitly stated. Only those active partial systems with links to the elements of the transformation system are considered. In the case of Earth-bound systems, it consists of the relevant parts of the geosphere (land and water), the biosphere, the atmosphere, the technosphere, and that part of the solar system that is responsible for climate. Time and space may conveniently be considered as parts of this environment for the transformation system.

The operators of a transformation system are mutually linked by various relationships, for instance by passing and receiving commands and data, energy and materials. These connections are implied in the model.

A particularly important link for technical systems, as for humans, is its connection to the *geosphere system*, see figure 3.3. Without this link, technical systems generally can not fulfill their functions. The linkage often takes place through an *action medium*. Thus a ship achieves the link by means of water, a flying aircraft by means of the air. Space travel has different links, where the environment of the solar system becomes more significant.

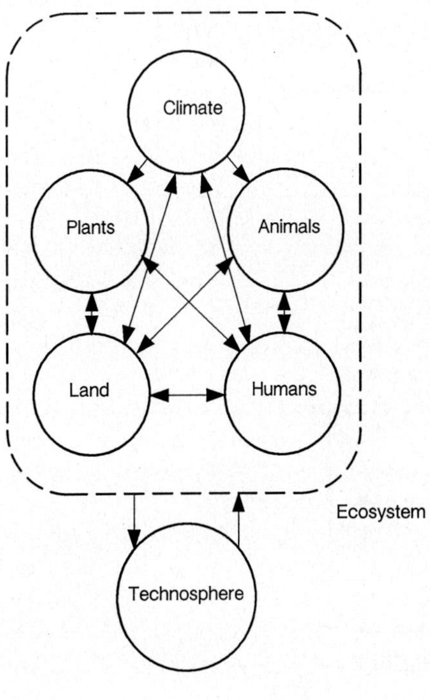

Fig. 3.3 Ecosystem and Technosphere

The *biosphere*, the living world, encompasses all living organisms and their living spaces. It comprises humans, animals, plants, bacteria, vira, and so on, and the locations and environments where they live.

Various *ecosystems* can be formed by different combinations of geo-, bio- and atmosphere. Mutual relationships exist in these that enable the processes which manufacture and transform the materials and energy that are necessary for life. The equilibrium of these ecosystems should be respected, and considered when developing transformation systems.

The *technosphere* encompasses all technical systems that have been or will be produced by humans within the space and time under consideration.

This broad view of socio-technical transformation systems is useful to introduce the theory of technical systems. Nevertheless, it presents a more complicated picture than necessary for chapters 4–11. For convenience, we will only consider the transformation process and its direct operators (the human system and technical systems) for the more detailed discussions. All other factors will be collected in the active environment, as shown in figure 3.4a, including the information and management systems. The simplified diagram, figure 3.4b, now allows us explicitly to

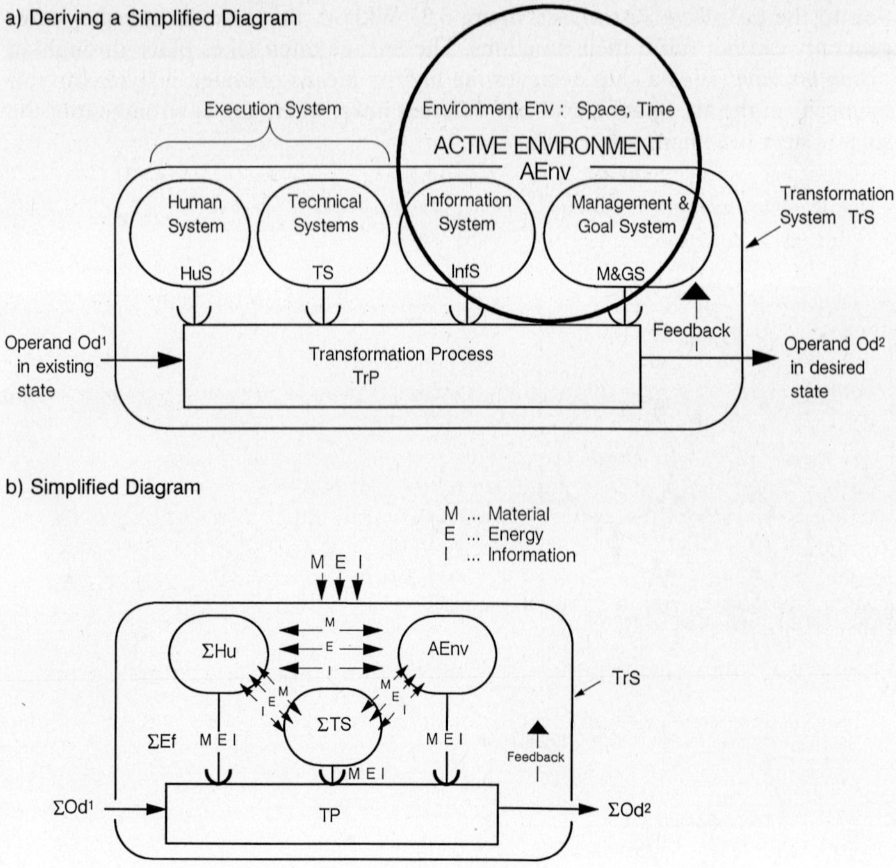

Fig. 3.4 Model of the Transformation System – Simplified

show the materials, energy and information supplied to and between the operators, and the material, energy and information nature of the effects that the operators supply to the operands in the process. This figure is the basis for the orientation diagrams at the title of each of chapters 3-12.

Restricting our view in this way helps us to focus on those parts of the sociotechnical transformation system that are under immediate human control. It assists by emphasizing the realizablility and effectiveness of a change to satisfy needs, and places the justification of the needs into the background (except in relation to those sub-processes of engineering design involved in evaluating the proposals and descriptions of the technical systems and processes being designed). Engineering has frequently been criticized for this attitude, but too broad a viewpoint generally inhibits action. Restricting the scope of our discussions to the process and its direct operators is thus justified.

3.3 Transformation Systems - Statements

This section summarizes and recapitulates the basic insights developed up to this point.

Statement 3.1 Human society has many needs, and these depend on that society's system of aims. Each individual aim (as an element of the system of aims) is a requirement for a particular object in a particular state. The objects can be regarded as the means of achieving the aims. If a desired state does not exist, then a new aim is evoked with a need to cause a transformation (to generate a state) that permits achieving the primary aim.

Statement 3.2 The changes of state are termed transformations, and the object that is being transformed is termed operand.

Statement 3.3 A transformation is an artificial process in which the changes are achieved by using natural phenomena.

Statement 3.4 The transformation is required (evoked) either because an unsatisfactory input state of the operand exists (when the choices concern only the transformation), or by defining an output state for the operand such that it provides the desired means to fulfill the aim (when both the input state and the transformation may be chosen).

Statement 3.5 Classes of operands in the transformation system are:

- biological systems - living matter, including humans, animals, plants, bacteria, vira, etc.
- materials - gases, liquids, solids, plasmas, etc.
- energy
- information - data, commands, news, signals, etc.

Statement 3.6 The transformation is performed by a technology that defines a system of partial transformations. In general, a number of technologies based on different technological principles are available to perform any transformation.

Statement 3.7 The transformations and partial transformations are realized by certain effects being exerted on the operand. Effects produced by the operators are the means of achieving the transformation. They are classifiable into material, energy and information (data and command) types.

Statement 3.8 The effects are delivered by various operator systems - humans, technical systems, information systems, management systems, and the active environment.

Statement 3.9 The transformation system is defined as the set of all elements that participate in the transformation. The model of a general transformation system is shown in figure 3.1b.

Statement 3.10 The effects are produced by the acting operators, by means of action processes and action chains within the operators that transform the inputs of the operators into their outputs.

Statement 3.11 A causal relationship exists between the transformations of the operand (as consequence) and the action processes (as cause).

4 Technical Processes

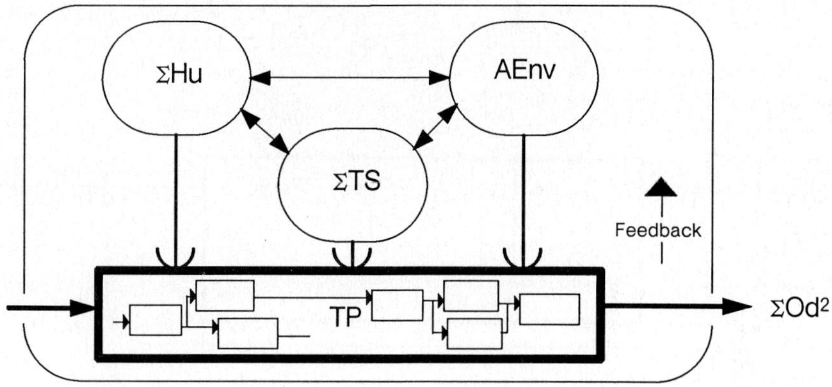

Of the large number of existing transformation processes as described in Chapter 3, those that involve technical systems among their operators are defined for the purposes of this book as *technical processes.* This chapter thus deals with a sub-set of all processes described by the general model, with emphasis on investigating those processes in more detail.

Technical processes are essential elements of socio-technical transformation systems, and they only exist to realize a (usually desired) transformation. All elements of the transformation system are brought together to perform a technical process.

The world is full of technical processes that continuously perform their duties. All people are usually simultaneous operators of a number of transformations, either directly to fulfill their needs, or indirectly to produce means (as a reserve or stockpile) to fulfill frequently recurring needs. The key word is *need*, see chapter 2.

What is the connection between a need and a technical process ? To restate the concepts from Chapters 2 and 3:

- An initial unsatisfactory state or situation exists, and people wish to or must change it.
- This state or situation forces a need to be formulated.
- The need may be fulfilled or satisfied by various means.
- The means with suitable properties to satisfy the need do not exist in nature.
- The existing state of the means must be altered in some way into a more desirable state.
- The change, the transformation of the means (the operand) from one state to a more suitable state, is accomplished in a technical process.

4.1 Model of Technical Processes - The "TP Model"

A technical process (TP) is an element of the transformation system, see figure 3.1. The process can be modelled, based on the relationships in the transformation system, as shown in figure 4.1.

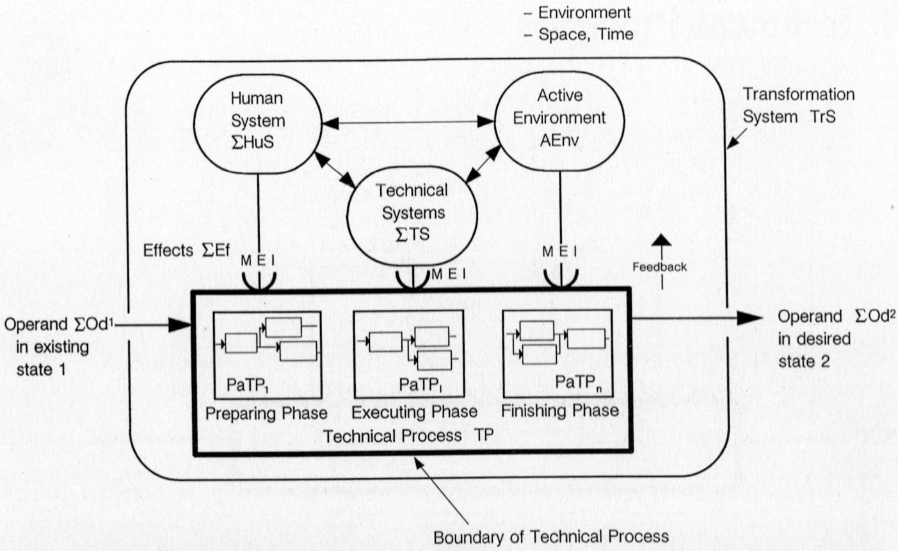

Fig. 4.1 TP Model – Model of the Technical Process

A specification of a technical processes describes the complete transformation of the operand, from "what" (entry state) to "what" (exit state) through "what" intermediate stages. This specification does not need to accurately specify "with what" (the operators), "how" (the effects received by the operand, and exerted by the operators), "who", "when" and "where". The "why" is provided by the needs.

The specification to every technical process should thus deliver suitable *answers to the questions*:

(a) What is the operand, and what are its states (entry, exit, and intermediate states)?
(b) Which transformations and technologies achieve the total transformation from input state to output state ($Od^1 \rightarrow Od^2$) within the framework of the existing conditions – including natural phenomena, societal limits and other restrictions or constraints?
(c) What are the effects to be exerted by the operators (classified according to categories of material, energy and information data or command types) that serve to realize individual transformations? – At least a rough definition is required, see point (f) of sub-section 4.1.1.
(d) Which of the operators of the process will exert (realize or produce) each individual effect?

The examples listed in figure 4.1 illustrate the form and content of answers to these questions. For a concrete problem, these items of information are best entered directly into a diagram (or other suitable graphical representation) of the technical process (TP), from which the contexts and relationships become clearer.

If we start from the premise that the output of the technical process (TP) is the desired state of the operand, then the *characteristic features*, including the following magnitudes and qualities of the TP, must be established:

Example	Operand			Transformation Procedure	Effects		
		State			Exerted in the form of material, energy or information by :		
	Type	In-1	Out-2	TgPc ... Technological Principle PaTP ... Partial processes	↓M	↓E	↓I
1	Material DIN St 50 SAE 1030	Raw material Form 1 Dimension 1	Shaft Form 2 Dimension 2	TgPc chip-forming machining	TS	Hu	Hu
				$PaTP_1$ feed, chuck	TS	TS	Hu
				$PaTP_2$ rotate workpiece	TS	TS	Hu
				$PaTP_3$ move tool relative to workpiece	TS	Hu	Hu+TS
				$PaTP_4$ measure		Hu	Hu
				$PaTP_5$ remove to bin			
2	Energy	Coal with chemical energy	Heat	TgPc combustion $C + O_2 = CO_2$ + Joule	TS	Hu+TS	Hu
				$PaTP_1$ store coal, ignite	TS	TS	Hu
				$PaTP_2$ burn, ensure air supply (O_2), remove heat, clean ash and slag		Hu	Hu
				$PaTP_3$ extinguish, clean			
3	Human	Faulty kidney	Functioning kidney	TgPc kidney transplant	TS	Hu+TS	Hu+TS
				$PaTP_1$ prepare human, achieve sterile conditions	TS	Hu	Hu+TS
				$PaTP_2$ remove faulty kindey	TS	Hu	Hu+TS
				$PaTP_3$ insert functional kidney		Hu	Hu+TS
				$PaTP_4$ observe			
4	Water (material)	In well	In tank	TgPc suction – pressurizing			
				$PaTP_1$ achieve partial vacuum, induct, guide	TS	TS	TS
				$PaTP_2$ pressurize, expell	TS	TS	TS
				$PaTP_3$ guide to tank	TS	TS	TS
5	Human (animal)	Place 1	Place 2	TgPc use transportation vehicle			
				$PaTP_1$ connect human in place 1 with vehicle		Hu	Hu
				$PaTP_2$ (human + vehicle) --> place 2	TS	TS	Hu+TS
				$PaTP_3$ disconnect human from vehicle in place 2		Hu	Hu

Fig. 4.1 (cont.)

- the exit state of the operand;
- the technological principle, based on the natural phenomena which are used to perform the transformation;
- partial processes and operations, which are established in broad terms by the technological principle, but the details of types of operation and their sequence must also be established;
- effects needed from the operators, which are derived from the needs for the operations;
- distribution of the effects between the operators (particularly between the human system and the technical systems), which must be suited to the assigned task and the problem situation.

These features represent the degrees of freedom available to a designer in choosing, formulating and optimizing the technical processes.

4.1.1 Some Explanations to the Model

The following explanations may be useful to clarify the general model of a technical process (TP), and to aid in formulating, composing and elaborating the duties of specific technical processes:

(a) The *principal stream* in the technical process (TP) changes the state of the operand. Multiple and interdependent operands are usually found and transformed in different ways within any TP, but are collectively termed "the operand". In each partial process and in each operation, the input and the output is the same operand, but in a different state. This homogeneity of the flow in the principal stream is an important feature of technical processes, and is reflected in the principles of conservation of mass, momentum, energy, etc. which are well established in the sciences. For example, this rule would be violated if one tries to describe as a technical process one in which the input is "fuel", and the output is "a journey" – the operands are obviously different.

(b) An output of a TP can only consist of those components that flowed into the process as its inputs. They appear at the output of the TP in a changed or unchanged state. A correct and easily reviewable balance of operands must be maintained. For instance, the balance is incorrect if the inputs are only specified as "coffee powder" and "water", and the expected output is only given as "a cup of coffee" – various other inputs and outputs must be stated to preserve the balance. If the input to a metal-cutting process is a piece of steel in semi-finished form, the output should strictly be a steel component and steel cuttings, not just the component.

This balance can in fact be performed with respect to each type of operand – biological and non-biological material, energy, and data and command information. In practice this is only possible once the transformation system has been fully realized (designed, developed, manufactured, installed, etc. and ready to be operated). During the conceptual phase of designing a technical process (TP), only a rough balance of the operand can be established, usually only for the principal flow of the transformation.

(c) In a rigorous interpretation, the technical process (TP) should contain only operations with operands. From a pragmatic point of view it is usually desirable to include in the TP various additional tasks such as preparing, cleaning, etc. These additional operations are referred to as being "evoked" by the process, particularly if they are directly connected with, or necessary to perform the transformation. In this way a more complete picture is obtained of the tasks connected with accomplishing the transformation, and a firmer basis for economic considerations is established.

(d) The model depends on an important agreement that each of the partial processes and operations in the model is assumed to contain not only the transformation itself, but also every auxiliary, propelling, directing, regulating, connecting

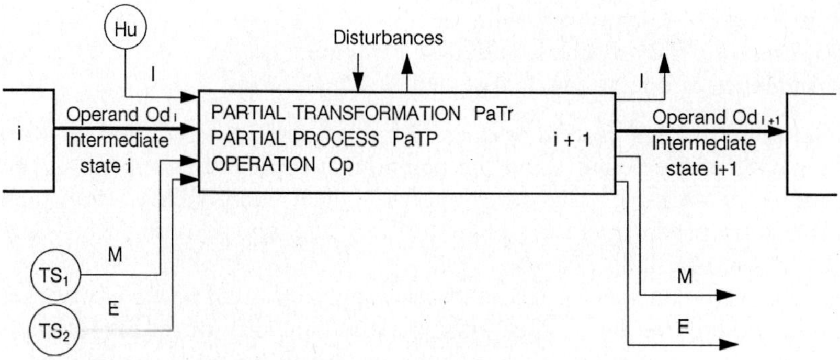

Fig. 4.2 Detail of a Partial Process (Operation) from the TP Model

and supporting process needed to perform that transformation (this will be explained in more detail in Section 4.4).

(e) Specifying the additional flows or inputs of material, energy and information which are needed to realize the transformation (see point (d) above) can be done in a relatively rough fashion. Figure 4.2 represents an "enlargement" or "window" of a detail part of a technical process - a partial transformation, partial process or operation. It also shows how the operators that deliver individual effects can be represented.

(f) Every process, partial process or operation progresses through three stages - preparing, executing and finishing (closure or conclusion). Each of these stages is associated with a group of questions to be answered by decisions which must be made during designing and/or operating a TP.

(g) The concrete process, partial process, sub-process or operation takes place within a certain space, and takes a certain time to accomplish. This information comprises all the physical data (temperature, pressure, humidity, dust content, etc.), all social and financial data, and a characterization of the time span, and should be contained in each TP model. Time is an important characteristic for technical processes, especially because all TP are dynamic in nature, and may be stable or unstable, growing, declining or oscillating.

(h) The description of transformations in the technical process (TP) may be rough or detailed, depending (1) on the purpose of the model, (2) the state of knowledge (including available information), and (3) the phase of processing (e.g. conceptual design) of the model. In any one model, the levels at which the individual operations are represented should be reasonably uniform.

(i) Every operation should be examined to make sure that it can be realized. If doubt exists, a request or contract for research can be issued in an attempt to deliver the necessary information.

(j) Checking and verifying a proposed technical process, and formulating alternatives and variants to this process, can be aided by the following questions:

- Can a particular partial process or operation be omitted ?
- Should an operation be added ?
- Is it possible or advantageous to replace one operation by another ?

- Can the sequence of operations be altered ?
- Can and should operations be separated or combined ?
- Can the sequence of operations be reversed ?

(k) The description of each process or operation must contain a verb and a noun, namely the nature of the activity, and the name of the operand as subject of the transformation, for example "transform energy", "heat component", "transport human". This form of statement is similar to the "function" as defined in value analysis [22, 84, 94, 144, 158].

The choice of verb can vary widely, and can signify different levels of abstraction. Since this description has a large influence on the choice of means for solving the problem, it is advisable to use a more general formulation, usually at a higher level of abstraction. Compare the difference between "saw off part" and "separate part". For added clarity, the verb may be capitalized, such as "HEAT component", see examples in [119].

(l) In any transformation, the operations are performed in a time-dependent sequence. When one operation is completed, and thereby an intermediate state of the operand has been achieved, then the next operation can be started. The flow of operand is roughly along a time axis. The operand can flow through the operations as a single object, as a batch of similar objects, or as a continuous stream. This concept is similar to the precedence relationships defined in critical path methods of planning [47, 96, 149, 197].

4.1.2 Conception of the Model

The model of the technical process (TP) as presented is only one of many possible configurations. The model can be constructed in various different ways, and the elements selected according to other rules within the framework of the possibilities. For instance, all the necessary propelling and controlling operations could be incorporated into the model instead of the energy and information flows. In the model presented here, the suggested restrictions in selecting transformation and working processes (operations) have the advantage that attention is focussed on the essential and recognizable aspects at this level of abstraction.

4.1.3 Application of the Model

The model of the technical process (TP) may be applied almost without restrictions to any technical activity. Even the human and sociological aspects of transformations can be included. The general interpretation of this model is useful in those circumstances where humans can and actually do participate in the transformation, and where the technical system can conventionally be classified as a "machine". The second statement implies that the technical system (TS) exerts a complete set of effects, and does not need the co-operation of any other TS to perform its duties. The model is thus applicable to any TS, for example, a spade, a plough, a tractor, etc.

This model of a technical process is less applicable to items such as a rear axle, a gear box, or a bearing in a tractor. As mentioned in chapter 3, those technical

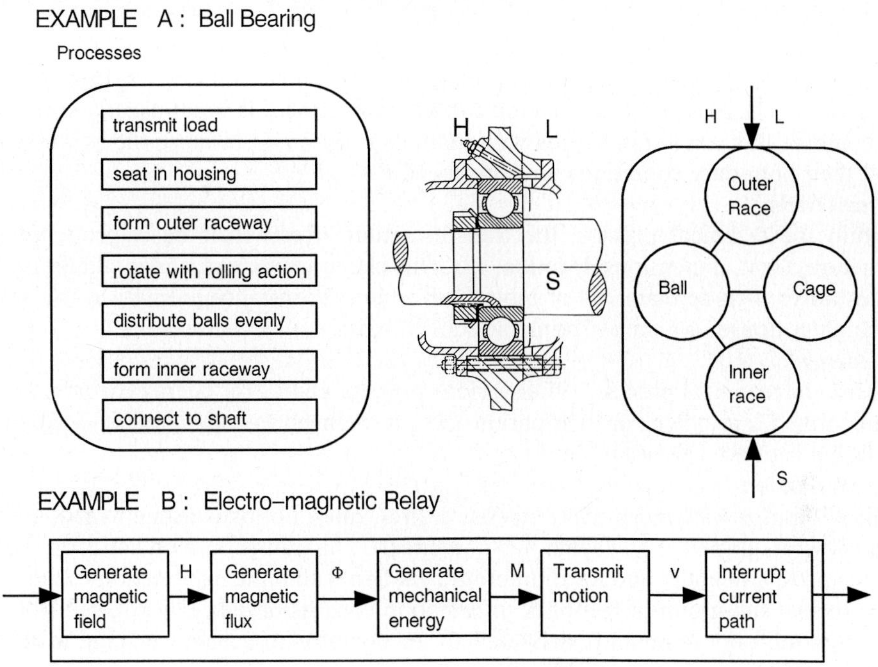

Fig. 4.3 "Technical Processes" (= Action Processes) Inside Technical Systems

processes that occur purely *within* a technical system (and acting without participation of a human being) are a particular kind of process. They are characterized by a particular purpose (that of providing effects) and their single operator (another TS). Consequently, the only factor that allows an engineer to formulate or propose variations or alternatives for solutions to the problem during the design process is the mode of action of the technical system - how it works. Figure 4.3 shows some examples of internal "technical processes" that occur inside technical systems. It would appear from this figure as if one technical system (e.g. the bearing ball) exerts effects (e.g. force) on another (e.g. the inner race), leading to a false conclusion that therefore the second system is the operand of the first. In order to avoid this implication, the types of processes that occur within a technical system are termed *action processes.* In general, a technical system should thus not appear as the operand in the process driven by that or another technical system (refer also to Statement 4.4), unless the object-system is actively being transformed (in which case, it should be labelled as the operand of the process).

4.2 TP Operands

The operand of the technical process is also the operand of the transformation system with which the TP is associated. We can distinguish four classes of operands, as follows:

a) **biological objects**
Such objects consist of living individuals or groups of human, animal or plant life-forms. Within the technical process applied to these biological objects, their state (e.g. sick → healthy) or their location can be transformed. It is not usually correct to combine this class of operands with inanimate materials, because the processes and their boundary conditions are very specific.
b) **materials**
Within the technical process, the transformation affects their basic properties, structure, form, dimensions, location, etc. This classification can under certain circumstances include materials of biological origin, for instance animals in the TS "slaughter house" which performs the transformation alive → dead.
c) **energy**
Within the technical process, either various types of energy (or energy carriers) are transformed into others, or their parameters are changed (e.g. $p_1 \rightarrow p_2$, $t_1 \rightarrow t_2$), or both kinds of change occur simultaneously.
d) **information**
This set comprises commands (requests, desires, rules, normative statements), and data (verbal, graphical and symbolic/numerical). The transformation concerns the form, quality, quantity and location of information within the information carriers. The special sub-group of feedback information consists of data interrogated from an operand, and commands delivered to an operator to cause a change in the exerted effects.

These four classes of operands are not independent. In general, no one class can be transformed without affecting the others. For instance, information cannot be moved from one location to another without simultaneously moving energy and material (at least in the form of subatomic particles). Nevertheless, it is useful to focus attention on the most important aspect and class of the operand for a particular process, and to give only passing regard to its other classes.

The state of the operand can be defined as a "vector" of its properties (or as a tensor, if this is considered more appropriate), in analogy with mathematical notation:

$$Od^1 = \begin{pmatrix} Pr^1_1 \\ Pr^1_2 \\ \dots \\ Pr^1_n \end{pmatrix} \qquad Od^2 = \begin{pmatrix} Pr^2_1 \\ Pr^2_2 \\ \dots \\ Pr^2_n \end{pmatrix}$$

4.3 TP Structure

The transformation to which the operand is subjected is often complex. The operand assumes step by step a number of intermediate states (which can usually be precisely specified), accompanied by progressive changes in various properties. The changes in the operand's properties can be considered as either continuous (e.g. heating of a workpiece) or stepwise (e.g. machining a shaft). Both types of transformation take place over a period of time, the property *time* takes on a different measure from beginning to end of the transformation. In fact, in a "stepwise" change, the intermediate transformation of state may be continuous (as in machining a shaft), but the transition states are of no interest in this context.

The simplest process is termed an operation, and consists only of work performed. Definitions of the term "operation" vary widely. For the purpose of this book, an operation is regarded as an uninterrupted part of a transformation (particularly unbroken in time), for instance characterized by:

- being allocated as an independent task to a worker or group of workers supported by one or more technical systems,
- being performed at one workplace,
- being concerned with only one operand.

The operation achieves generally a desired change in a single property of the operand.

For example, the operation "turning (machining)" produces bodies of rotational symmetry, "milling" produces straight grooves, "hardening" produces a harder surface. Figure 4.4 shows three operations involving changes in two properties, Pr^x_1 and Pr^x_2. Property Pr^x_1 changes stepwise by ΔPr^x_1 in the partial operation $PaOp_1$, during which the operand has been subjected to the effect ΔEf_1. The transformation $Pr^1_2 \rightarrow Pr^2_2$ is a continuous change during the third operation, whereby property Pr^x_1 remains unchanged. Both properties are unchanged by the second operation, transportation.

This definition still allows different interpretations depending on the viewpoint. An engineer may define an operation as "machine a shaft" with respect to manufacturing a complete gear box, but in production planning for machining that

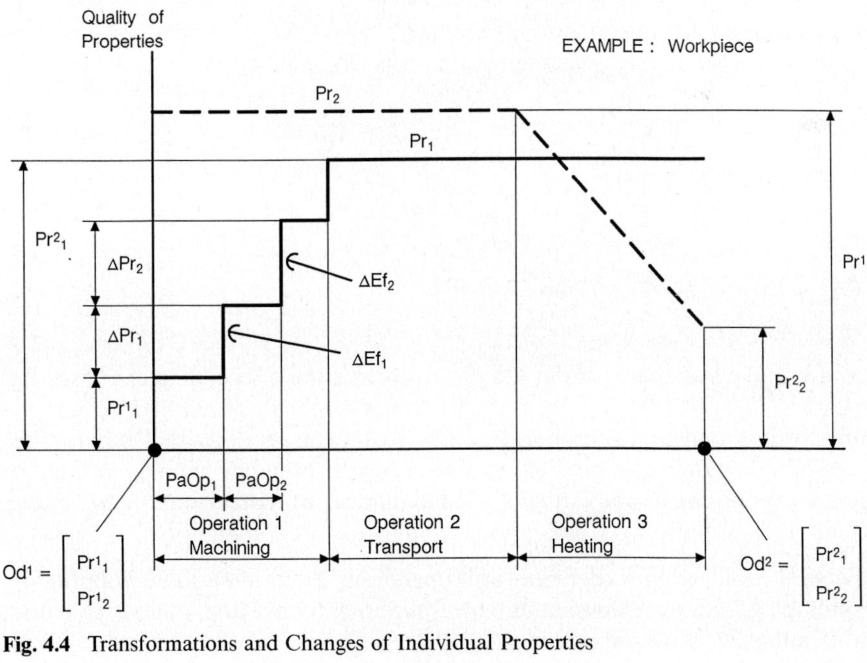

Fig. 4.4 Transformations and Changes of Individual Properties

shaft an operation may be defined as "mill a keyway", or "turn a snap ring groove".

The totality of operations and their sequence is termed a *working process*, and is determined by the applied technology. The choice of technology (a collection of individual technologies for each operation) depends on the given requirements, and also on the operators that may be used to drive and guide the transformation process. Important features in this choice are the available technical knowledge, including knowledge about properties of operands and characteristics of transformations, and especially knowledge of the technical systems that may be employed to realize the transformation. A working process is subject to change with time, depending on progress and the acquisition of insights and knowledge.

The structure of a technical process (consisting of the elements "operations" and their relationships) depends primarily on the technology, as shown in a general form in figure 4.5. A more complete specification of the process structure

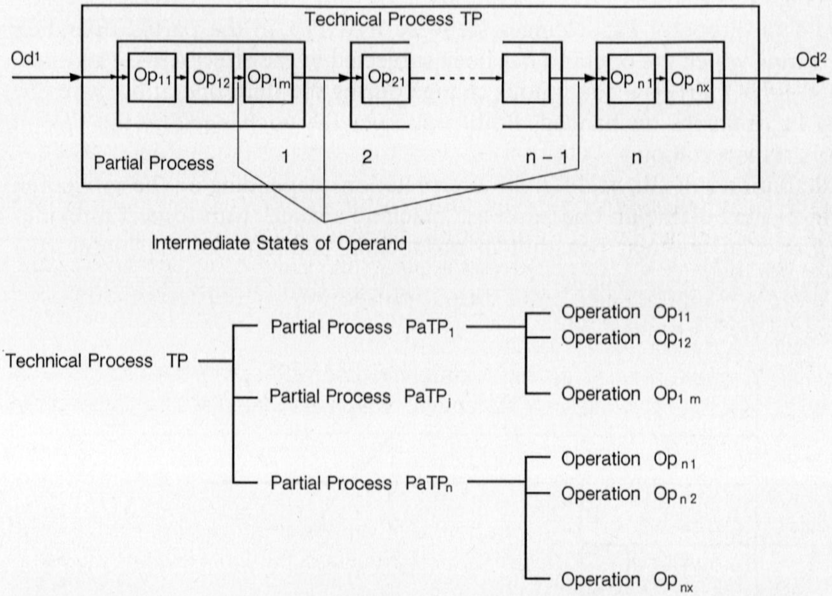

Fig. 4.5 Structure of a Technical Process

must also make statements about the operators and the effects they exert, see figure 4.6.

Sub-dividing a technical process into its constituents as described above raises the question whether it is possible to find a degree of subdivision such that the structure can be seen as consisting of a small number of repetitive *basic operations.* The answer is well known from various technical areas and authors:

- chemical engineering recognizes "unit operations"; comminution (chopping up, grinding), partition, agglomeration, transportation (conveying, pumping), oxidation, reduction, polymerization, etc.;

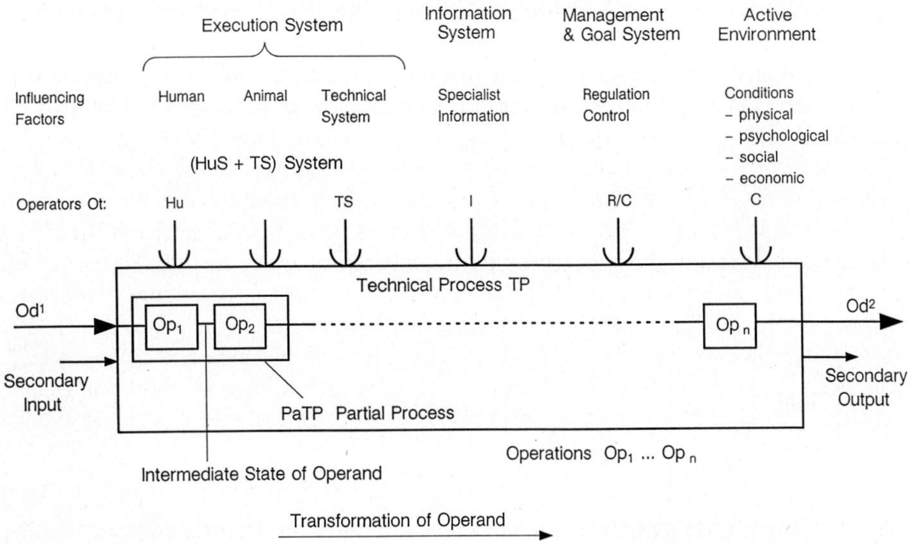

Fig. 4.6 Transformation Needs Various Effects Exerted Onto the Operand

- Roth [187] lists operations of storing, transmitting (conducting, changing), transforming, combining (joining - summative, parting - distributive);
- a more general set consists of processing (changing the internal structure), manufacturing (changing the external form), transporting (changing location), and storing (changing time) [119].

A further question is concerned with finding variants to the process structure. This may be resolved by the following statement, which will be discussed in a more general form in Chapter 7:

PROPOSITION 4.1

The transformation (and thus the behavior of the system or process) $Od^1 \rightarrow Od^2$ may be realized by various structures of processes (and therefore by various technologies).

4.4 Received Effects, Produced Effects, Effects

The operand must be acted upon by suitable effects in order to achieve the transformation resulting in the desired properties at the output of the process. Things must be heated, mixed, have material removed, transported, healed, and so on. The effects that are needed to transform the operand in the process are termed *received effects* (from the viewpoint of the operand), working effects, or transforming effects (which need to be delivered by the operators).

The working effects are achieved using known phenomena as recognized by the natural sciences. Heating, i.e. increasing the temperature of a component, is based on the insights about physical phenomena, how heat can be generated from various forms and carriers of energy, and how it can be transmitted to a component.

The total effect (Ef) needed to achieve a transformation ($Od^1 \rightarrow Od^2$) consists of partial effects (PaEf) that change certain properties, e.g. $Pr^1{}_i \rightarrow Pr^2{}_i$. This can be expressed as:

$$Pr^2{}_i - Pr^1{}_i = f(PaEf_i)$$

and the total effect is the sum of the partial effects:

$$Ef = \begin{bmatrix} PaEf_1 \\ \dots \\ PaEf_i \\ \dots \\ PaEf_n \end{bmatrix} \qquad Od^2 - Od^1 = f(Ef)$$

The working effects alone are not sufficient to realize the transformation. Various other effects are required, and usually present for every process; they are normally not mentioned because these additional or accompanying effects are generally performed by human beings. Typical classes of these *additional effects*, illustrated by examples from the technical processes of "machining", are:

- auxiliary effects, e.g. centering, tool lift, cooling, lubricating
- propelling effects (material or energy converting), e.g. main spindle drive, tool slide drive
- regulating and controlling effects, e.g. tool setting to dimension, measuring, tool mounting, machine starting
- connecting and supporting effects, e.g. machine system as a unit, housing and frame of machine.

Note 1: Propelling, regulating and controlling, connecting and supporting may also be regarded (in an extended view) as "auxiliary effects" to the main working effects. They are distinguished here for pragmatic reasons, because these additional effects are almost always present in any technical process.

Note 2: The words "regulating" and "controlling" used in the description of these additional effects are generally considered to be almost interchangeable. The distinction implied by the dictionary definitions and used in this book are that *regulating* provides the rule and the governing input to this partial process, and *controlling* senses the output and applies corrections to return the process to its desired state. It is useful to make this distinction so that both the function of setting a standard (an executive function) and the function of complying with it (an execution function) are recognized. A direct analogy can be drawn with legislation and law enforcement.

Figure 4.7 contrasts two models of the transformation system known as a "lathe". Part a shows a conventional diagram as shown in the literature, in which the workpiece is modelled as a direct input to the transformation system, the tech-

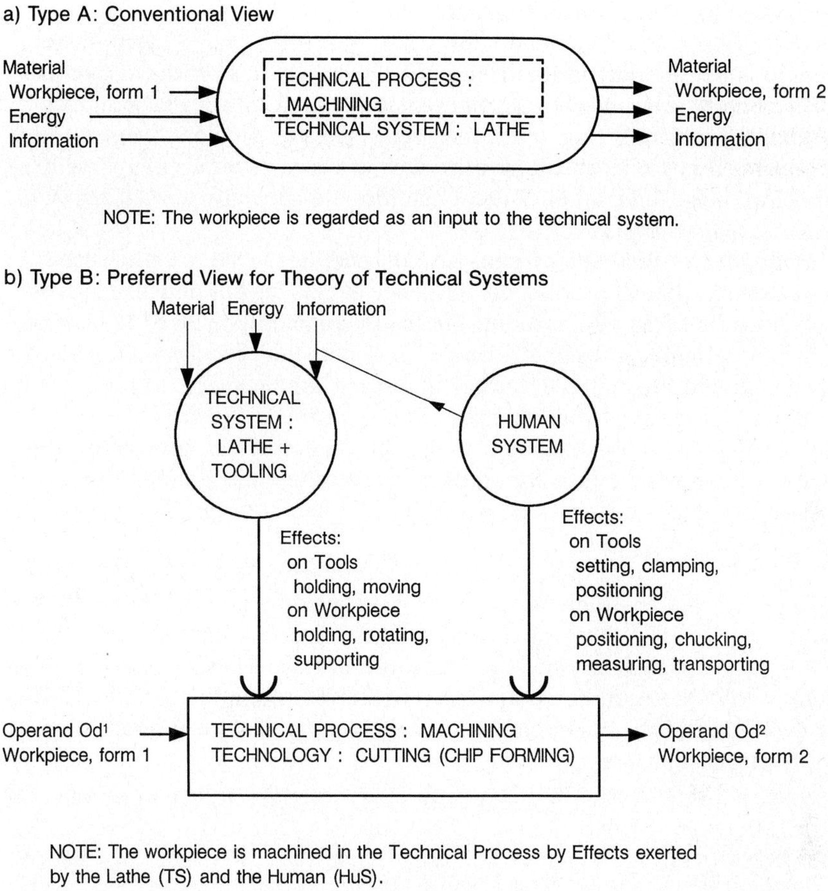

Fig. 4.7 Formulation of the Transformation Model

nical process of turning and producing the finished workpiece is hidden inside the system.

The model preferred in this book is part b of this figure, which shows the process of machining, with the technology of "metal removal from the workpiece by chip formation at a sharp tool edge", transforming the operand, the workpiece. The effects supplied by the technical system and its parts, and by the human machinist are clearly presented and allocated to the appropriate operator.

The same effects (motions, forces, decisions, etc.) supplied by the same operators can be used to drive and guide a number of different processes. For instance, with a different form of tool (and technology), a lathe (the same technical system) can be used (by the same human operator) to coil springs (a different process). The nature of the process, its operators, the exerted effects, and the technology are thus clearer in the preferred representation and model, consistent with the theory of technical systems as presented in this book.

4.5 TP Secondary Inputs and Outputs

The operand is not the only input or output from a technical process. In order to realize a technical process, various other input quantities are needed, such as energy, auxiliary materials (e.g. lubricants, catalysts), etc. Similarly, various other outputs appear from the technical process, such as waste materials (chips, smoke), heat, vibration, noise, and so on. These quantities are generally termed *secondary inputs* and *secondary outputs.*

The importance of secondary inputs and outputs has to be acknowledged. In order to realize a technical process, the influence of existing but undesired secondary inputs must be taken into account. These are commonly referred to as *disturbances.* While designing a technical process, all potential disturbances from the environment should be analyzed, and, if necessary, arrangements to monitor the critical levels of important disturbances should be made. Such is the case for example for high precision machine tools, the manufacturer prescribes upper and/or lower limiting measures for ambient temperature and humidity such that the machine can perform its processes within the tolerances specified in the performance data for the machine.

Similarly, the secondary outputs must be carefully considered, because humans (and their society and environment) are now threatened by some of the outputs of technical processes. The local environment of the process may be subjected to undesirable changes, such as *pollution.* The cause of this recently perceived threat is that up to now the secondary outputs have received too little attention. Consequently, protection of the environment has become one of the important requirements of a technical process.

4.6 TP Operators

An important question with respect to the transformation is how those effects needed to transform the operand can be achieved to drive the technical process. Let us consider the operations for hardening a workpiece. In a traditional way, a worker can grasp the heated workpiece in a pair of tongs, remove it from the furnace, dip it into and remove it from the coolant, and place it in a receiving bin. A different way would be for the worker to place the workpiece onto a conveyor, allow it to travel through a continuous furnace to heat it, bring it to the coolant, remove it, and let it drop into a prepared bin. In the first case the human operator has exerted almost all the auxiliary, propelling, regulating and controlling effects, including transporting, with the technical systems delivering mainly heating and cooling. In the second case the work performed by the human operator is limited to the operations of preparing and starting the automatic hardening line, material handling (which could incidently be transferred to a programmable robot), checking and maintaining.

The total of effects required to act on the operands in the technical process is realized by the *execution system* of operators, consisting of humans, plus technical systems, with different levels of participation from each. Note that for completeness we should also consider animal and vegetable life-forms as participants in the operator system, because there are still applications where animals (and sometimes plants) appear as operators - horse-drawn ploughs in alpine agriculture, horse- or dog-drawn carts and sleds for specialized transportation, bacteria in fermentation and other bio-technology applications, etc. The complete system of operators should therefore be stated as consisting of "humans, plus animal and vegetable life, plus technical systems".

Some other operators are important for the overall result achieved by a technical process. They influence the effectiveness and efficiency of the process. Sometimes they may even obstruct or prevent realizing the process. These operators must therefore be taken into account, and their range of influence recognized, particularly where accidental failures may occur. These operators constitute an *active environment*, and include:

- technical information about natural phenomena, engineering, science, processing and manufacturing technology, and others. This total of available information with respect to existing technology and technical systems at any given time represents the *state of the art*;
- process management, including regulating and controlling personnel, coordinating activities, checking, procuring materials, etc.;
- conditions of the environment:

  physical - temperature, pressure, space, dust content; psychological - relationships between human beings; social - state and interest in society at a given time; financial - monetary means to accomplish the process, especially capital funds, running costs, cash flow, inventory, and so on.

The results of the process, the desired output state of the operand, is consequently determined not only by the execution system "human being - technical system", but also by these other operators, collectively shown as the active environment (compare figures 3.1b and 3.4).

These factors can be viewed as flows of materials, energy and information. This interpretation does not present any difficulties with factors such as technical information, methods or management, because these can be regarded as forms of information. It is also consistent with the idea that information can exist and be used in two basic forms, commands or data (section 4.2).

4.7 TP Characteristics and Evaluation

Usually a technical process (TP) is characterized primarily by its output, that is the properties and quality of the operand at the process exit. The technical process fulfills the requirements completely if all the desired properties of the operand have been achieved. It can happen that some properties were not correctly recognized before or during the design phases (a circumstance which should be avoided wherever possible). Such unanticipated properties may drastically affect the achieved results of the process (in contrast to those anticipated from an assessment prior to operation). In particular, they can make the difference between a product (operand in the output state) being or not being marketable and acceptable.

Another way of characterizing a technical process is by the level of its operators:

- the level of the technology,
- the level of the employed workers and technical systems (the means)
- the level of the specialist technical information, management and conditions.

The causes of failure may be found using these characteristics, if the technical process does not (completely) fulfill its duty of transforming the operand.

These levels are measured relative to the *state of the art* at that time, and may reflect either any deficiencies in the state of the art, or its interpretation, or in the realized technical processes. The state of the art is not constant, it generally improves with time. It is also at any one time subject to interpretation, and depends on how well the current technology is accepted in the circumstances. The state of the art existing in research establishments is generally much higher than for the market-leader companies, who in turn perceive a higher state of the art than general industry.

Two summarizing and guiding numbers may be used to characterize the process and provide a means of comparing that process to another. These are commonly known as efficiency and effectiveness. According to Drucker [55], efficiency is "doing things right", and effectiveness is "doing the right things".

The *efficiency* of a process is the ratio of the amount of a single quantity (e.g. of energy) appearing in the useful output to the quantity of the same kind entering the input to that process.

The *effectiveness* of a process is defined as:

$$\text{Effectiveness} = \frac{\text{Benefits of the process (yield, income)}}{\text{Expenditure for the process}}$$

It generally gives an aggregate of benefits and expenditures, usually expressed in some common measure, such a money. This measure of effectiveness is roughly equivalent to a "cost-benefit ratio" [83, 148]. It is also similar to the definition of "value" used in value analysis [22, 84, 94, 144, 158]. As usual, a single measure such as effectiveness, or cost-benefit ratio, can only provide an indicator to the nature of the process for the purpose of evaluation. Such indicators must be interpreted with care, because they omit, hide, summarize, simplify or ignore many features that may be important. The benefits of the process are determined by the effects exerted on the operand, particularly by the changes in properties of the operand. In many cases some of the secondary outputs of a process may be usable – as by-products or by re-cycling – and can increase the benefits of the process.

The expenditures represent all items of the transformation system, whether they concern operands, secondary inputs or operators. They include the costs of operands (e.g. raw materials), energy, auxiliary materials, wages and salaries, social costs, depreciation, maintenance, overheads, special costs, etc.

The two concepts of efficiency and effectiveness may be illustrated by considering a filter for a fluid medium containing solid particles. The efficiency of that filter is usually expressed as the ratio of particles passing through the filter to those entering. As a filter is used, particles are retained in the filter, and slowly accumulate. The filter will then gradually retain a larger proportion of all particles, and the statistical distribution of particle sizes retained will drift towards smaller particles. Its efficiency gradually increases, until in the limit it will block passage of all particles. The effectiveness of the filter depends on an additional quantity, namely the amount of fluid passing through the filter. Let us assume that the fluid contains a constant amount of contaminating particles, and that the available pump power is limited. Then, as a filter is used, the pressure drop in the filtering medium increases, and the flow rate of fluid through the filter progressively decreases. Initially, the effectiveness of the filter will increase, as the filter removes more particles from the fluid. When the flow rate is sufficiently reduced (or in the limit is completely blocked), the fluid can no longer be adequately filtered, the effectiveness of the filter system reaches a peak, and then drops gradually towards zero.

In order to uniquely determine and characterize a technical process it is necessary to consider the measures of all the properties related to that process. These comprise a number of characteristic indicators, tabulated in a usable form in figure 4.8. They are divided into three classes: technical, economic and planning.

Characteristic Group	Examples of Characteristics		
	Technical	Economic	Planning
Operand	Material, dimension, form, surface finish, pressure, temperature, and other parameters	Price, costs	Production quantity, delivery deadline, supplier
Technology (Working Procedure)	Specification of operations, sequence of operations	Costs	Working time, worker, workplace
Worker	Technical knowledge, experience, personal character properties	Labor costs	Work period hours, number of workers
Technical System	Functional properties, operational properties, appearance properties, ergonomic properties, distribution properties, and others, see fig. 5.1	Price, Operational costs	Delivery deadline, quantity, supplier
Technical Information	Information index, information source	Costs	Deadlines, examiner, workplace
Regulating and Controlling Processes	Organizational system, planning system, remuneration system, administration	Costs	Deadlines, examiner, workplace
Conditions of Environment	Physical – space arrangement, – space requirements, – temperature – humidity – light – noise	Costs	Deadlines, worker, workplace
	Psychological – working climate		
	Social – political situation	Economic situation	

Fig. 4.8 Characterization of Technical Processes

		ENERGY TRANSMISSION	Mechanical	Hydraulic	Electrical
Criteria	1	Distance	–	+	++
	2	Control of transmission	–	+	+
	3	Safety	equal		
	4	Efficiency	+	–	–
	5	Accessibility of components	–	+	++
	6	Adaptability for design	–	+	++
	7	Price of execution	–	–	+

EXAMPLE : Energy Transmission Processes

Fig. 4.9 Evaluation of Technical Processes

Classifying Aspects	Classes of Process	Comment
Operand	– processes that treat material – processes that treat energy – processes that treat information – processes that treat biological objects	

Phenomena to achieve Working Effects	– physical processes
	  – mechanical
	  – electrical
	  – thermal, etc.
	– chemical processes
	– biological processes
	– combined processes	

Working Effects	– transport process – sorting process – comminution process – manufacturing process – assembly process	
Type of Propelling Effect	– processes with manual work – processes with animal power – mechanized processes	
Type of Regulating and Controlling Effect	– human regulated and controlled – automated – computerized (programmable) – hybrid	
Level of Complication of Process	– operation – partial process – complicated process	
Relationship between Input and Output	– combining process : number of inputs > outputs – dividing process : number of inputs < outputs	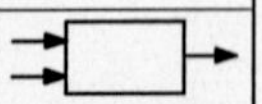 
Progress of Transformation	– continuous process – discontinuous process (discrete, batchwise)	

Fig. 4.10 Systematics of Technical Processes: Some Classes of TP

The formal procedure of comparing and evaluating technical processes is similar to that for object systems, which will be treated in chapter 8. A rough comparison of energy transfer processes by a "plus-minus" method provides an example of such an evaluation, and is shown in figure 4.9. Alternatively, a point-rating scheme could have been used, probably with the same criteria.

Here it is useful to emphasize some of the social and economic consequences that follow from engineering decisions. Any decisions or recommendations from an engineer concerning a technical process is

(a) generally the prerequisite for employment of humans,
(b) influences the social and economic situation in a region or country, and
(c) has a number of other repercussions, e. g. on the environment.

4.8 Systematics of Technical Processes

Technical processes can be categorized from various viewpoints. Figure 4.10 illustrates some of these viewpoints with their classes of process. One of the most important characteristics for placing technical processes into these classes is the type of operand - biological and non-biological material, energy and information. These four types of operand may be mapped against the four basic types of transformations (of structure, form, location and time, see section 4.3); the resulting sixteen basic types of transformation processes are shown in figure 4.11.

TRANSFORMING PROCESS		TRANSFORMATION of structure	of form	of space coordinate	of time coordinate
		PROCESSING	MANUFACTURING	TRANSPORTING	STORING
Operand	Material M	M convert (process)	M transform (form, shape)	M transport	M store
	e.g.	iron ore → steel	scantling → workpiece	in storage → on bench	workpiece in bin
	Energy E	E convert	E transform	E transport	E store
	e.g.	hydraulic → electrical	50 000 V → 220 V	power station → consumer	in accumulator (battery)
	Information I	I convert	I transform (translate)	I transport	I store
	e.g.	graphical → digital	German text → English	news sender → receiver	speech on magnetic tape
	Human Hu	Hu convert	Hu transform	Hu transport	Hu store
	e.g.	sick → healthy	hand → prosthesis	in London → in Toronto	at home

Fig. 4.11 Basic Technical Processes

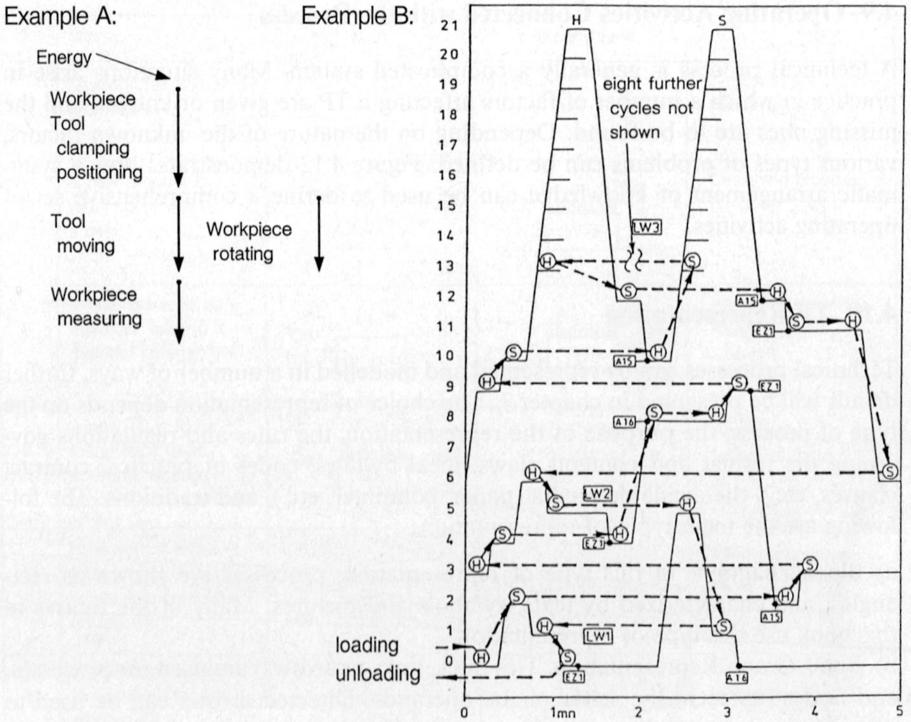

Fig. 4.13 Representation of Processes: Examples

is interpreted as "start by performing operation **A** (e.g. turning a shaft), then check that condition **p** has been achieved (e.g. ∅ 90 H7). If the condition is satisfied, then perform operation **B** (e.g. milling a keyway into the shaft). If it is not satisfied, then follow the arrow belonging to **p** back to its start, i.e. repeat operation **A**.

4.11 Special Theories of Technical Processes

The considerations up to this point have concentrated on a general model that is applicable to any technical process. Processes have a hierarchical nature similar to that of systems (see chapter 3), and many of the typical classes and families of processes known to man have been studied in the engineering sciences. Without invalidating the general model, such specializations as thermodynamics, fluid mechanics, manufacturing technology, chemical processes, controls, etc. provide opportunities to enter into the details of individual processes. The resulting *special theories of technical processes* are more specific, less general, and at this lower level of abstraction include much more detail information, and more accurate models of the phenomena. Similar considerations for technical systems will be discussed in more detail in chapter 12.

4.12 Technical Process - Statements

The general knowledge about technical processes can be summarized in the following statements.

Statement 4.1 A technical process (TP) is a special form of transformation process in which technical systems are used by humans as artificial tools. Consequently, all statements about transformation processes as listed in section 3.4 are true for technical processes.

Statement 4.2 The structure of a TP consists of a set of partial processes or operations, and a set of intermediate states of the operand.

Statement 4.3 Partial structures can be produced for the three phases of preparing, executing and finishing (closure or conclusion) to a TP.

Statement 4.4 A process in which the technical system itself is the operand (instead of a separate operand that is subject to transformation) is only acceptable in exceptional cases, and for pragmatic reasons.

Statement 4.5 Each transformation process (the work process that transforms the operand) requires for its performance an additional set of processes: auxiliary, regulating and controlling, material and energy converting (propelling), and connecting and supporting processes. The need for these processes is evoked by the main working process. The representation of a TP includes (by convention and agreement) all these evoked processes.

Statement 4.6 When considering a TP, its environment (and space) and the time must also be considered and specified.

Statement 4.7 Secondary inputs and outputs (including disturbances, waste and pollutants) belong to the TP-system and must be considered.

Statement 4.8 Even though technical systems are the most important operators of a TP, all other operators must be investigated.

5 Technical Systems

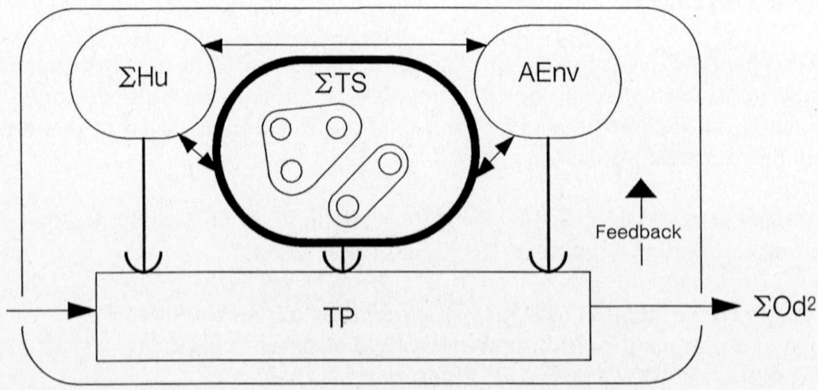

The best-known elements of transformation systems are the technical systems (compare figure 3.1). Chapters 5–11 of this book focus on technical systems.

As discussed in chapter 3, these technical systems (together with human beings, the information and management systems, and the active environment) exert onto the operands the effects that are necessary to accomplish the desired transformation. Technical systems are the principal means by which the transformation is achieved.

Engineering is recognized by its technical creations, just as art is recognized by its artistic creations. Individual branches of engineering have generated products that are practically unlimited in numbers, quantity, and variety. Particularly machine systems, as produced by mechanical engineering, contribute significantly to the variety and number of technical means. An immense set of machines, appliances, devices, armaments, means of transportation, jigs and fixtures, tools, and others exist to serve humans, or damage them, or stand idle, depending on how well their use and operating conditions are understood. Mechanical engineering obviously does not stand alone, most of these items are products that combine knowledge and experience from more than one branch of engineering. Modern systems are becoming increasingly multi-disciplinary.

In this chapter, we show how this variety and number of devices can be understood in some unified manner. We need to find order in our perceptions of the general features of technical systems, and particularly machine systems, and to establish general laws that are valid for all technical systems. These questions, although they appear to be self-evident, have not been adequately addressed by either the existing sciences or by philosophy.

5.1 Nature of Technical Systems (Habitude)

If the technical system is to be the center of attention in our observations, then some key questions need to be answered concerning the general nature of technical systems, and in particular: what is their purpose? what is their mode of action? what are their constituents? and what intermediate and final states does each TS attain?

All technical systems are (without limitation) members of the causal-mechanistic world picture, they follow the principle of causality. Every happening in the TS is the consequence of one or more causes, and is simultaneously the cause of one or more other consequences (refer also to section 3.1, and comments related to the choice of words in appendix B).

This does not necessarily mean that we need to know and fully to understand the causes of every happening, or that they need to be measurable. In many cases the causes are so complex, with a multitude of interactions, that humanity cannot fully comprehend them and uniquely assign a cause to each consequences. This is especially true where a human or other living element participates in the chain of events. Nevertheless, within the definition of causality, most technical systems are fully deterministic. Randomness appearing in the measurable causes and consequences can readily be accomodated by statistical methods. All parameters are probabilistic, with a lower bound set by Heisenberg uncertainty. Control of variance in technical systems is thus limited by natural phenomena and their inherent variances. The achieved variance can be influenced by designing to approach the natural limit.

Historically, in the development of philosophy, the principle of causality appears more or less as a strict association of one event to another. It is a constant subject of argument and controversy. Such arguments are not, however, applicable to engineering and especially technical systems, but are more related to the "machine theory of life" and similar questions [191].

Under closer observation, all causes have three major components [159, 172, 173, 191]:

- conditions under which something happens;
- internal actions through which it occurs;
- a trigger that initiates the immediate occurrence.

The consequence arises *when, always when, and only when* the causes occurred or are simultaneously acting. "When" and "always when" is analogous to the sufficient conditions for the cause-consequence chain, and "only when" is analogous to the necessary conditions.

These assertions rest on innumerable observations of nature and human experiences, and have been collected, categorized and modelled as the phenomena studied by physics, chemistry, biology, etc. The observations form the basis of human knowledge, and their ordering, categorizing and interpretation is the main task of science. Generalizing from these experiences has produced the principle of causality, not as an inviolable law, but as a useful and powerful guideline.

In contrast, humans are goal-oriented and want to make things happen (as purposeful events). From the viewpoint of the principle of causality, the human tries to produce an effect exactly when and where it is needed. We are concerned with *finality*, focussing on the consequences, results or outcome of a set of causes through which humans attempt to give a more "determined" form to their life, as opposed to causality, which implies a focus on the existing causes. Nevertheless, humans are forced to use the physical, chemical, and where possible biological events and phenomena to find cause-consequence chains of which the consequences are more useful to themselves, considering the technical-scientific requirements.

Using the idea of causality, humans employ a suitable systems of causes in deliberate cause-consequence chains (action chains) that will, when initiated and triggered at the desired time, deliver the needed effects because of causality .

This described reversal of causality (which we have labelled "finality") occurs in engineering when the technical systems are being designed. It consists of anticipating (conceiving) the causes that will realize the necessary causal systems, which in turn will produce the desired consequences, as the effects delivered by the TS. In the process, certain limitations, boundaries and feasible spaces must be respected (particularly concerning the active environment), and the internal conditions and triggering procedures for the humans and the technical systems must be created.

The above excursion into philosophy was necessary to better understand the nature of technical systems. It permits a deeper discussion of their purposes, actions, constituents and states.

As in the biological sciences, the nature of technical systems can usefully be classified with respect to different aspects perceived in them. One of these relates to the tasks to be performed by the whole system (including its purpose), and the effects it should deliver (the functions), without reference to its internal workings. A second aspect considers the internal processes (the modes of action) and the necessary groupings of points, surfaces, volumes, etc. that can realize those functions (the organs or function carriers) without considering how they are interconnected. A third aspect considers how the system can be manufactured and assembled from individual physical parts (the components). A fourth aspect refers to the life stages of a TS, the states of existence and operation that the system achieves and passes through. We make no apology for using some of the insights and terminology of a much older science - biology - to guide this discussion, as its utility will become apparent in the following sections and chapters of this book.

5.1.1 Purpose

Let us return to the model of a transformation system, figure 3.1b. It shows that technical systems should exert planned and goal-oriented effects on the operands of technical processes. The tasks are holding, moving, storing, buffering, heating, connecting, enabling, dividing, sealing, driving, regulating, controlling, and a host of other effects that eventually serve to satisfy human needs.

These actual abilities of the technical system are termed its functions (compared to its intended abilities, its purposes). Unfortunately in this context the ability of a technical system "to convert its input quantities into relevant output quantities" is also defined as its ("technical") function (especially in the German literature, compare [100]).

Designating the tasks to be performed by the TS as its output effects, or its purpose function, is more unequivocal. These output effects (as an aim to be achieved) are related to the "technical" function as the means for achieving that aim. The difference between these two terms is frequently obscured in design discussions, especially with respect to technical systems of low complexity, and this tends to hinder communication. Ropohl [185] has chosen the terms "teleological function" to designate the purpose function, and "descriptive function" for the technical function.

In this book, the term "function" to designate an output task of a TS is avoided, and is replaced by the term "effect". The word "function" is retained to name the "internal tasks" of a TS, mainly because it is a commonly used term in such contexts as "function structure" and "function carrier", as will be shown in sections 5.3 and 5.4.

The purpose of a technical system is thus represented by the system of its output effects. This system of effects is frequently referred to in the literature as the "total function of the TS" [14, 167], which is acceptable and applicable for some families of technical systems, but leads to errors with respect to others. This question receives further consideration in section 5.3.

5.1.2 Mode of Action

Technical systems use recognized natural effects to fulfill their purpose. These include lever effects, gravitation, thermal expansion, electro-magnetic force, and the outputs of many other physical, chemical and biological phenomena.

The constituents of a technical system are fitted together in such a way that their concerted action (the TS-internal process) drives and guides a desired events, and thus achieves the required external effect. If a gear wheel is rotated, it forces the mating gear into rotational motion, and by the mode of action of the gear pair the rotational speed and direction are changed. As second example, consider a fluorescent light tube: electrical current produces heat in the filament, which (because of the applied voltage) emits a stream of electrons from its oxide layer into the gas mixture that fills the tube. Aided by a small amount of mercury vapor, the electron stream causes the gas atoms to emit ultra-violet radiation, and this in turn is converted by a layer of fluorescent material into visible light. Thus the output effect, light, is achieved.

In order to obtain a certain result (i.e. an output effect), various phenomena are linked together into an action chain, in such a way that an input quantity is converted into an output quantity. This chain describes the mode of action of the TS (and thus, in common language "how it works" or "how it functions").

The *mode of action* describes the way in which the inputs to a technical system are converted into its effects (its outputs). This internal conversion can be described in two significant ways. One description uses the technical system's internal functions (tasks, what it does) and is represented in a "function structure of the TS". The other uses the means that execute the functions (the organs, or function carriers), and the same mode of action is described and represented in an "organ structure of the TS". According to the degree of complexity of the system, the second representation can use organisms, organs, or partial organs.

Each of the organs evokes a number of additional functions, which need to be fulfilled in order for that organ to be able to fulfill its duties. These additional functions are realized by organs of lesser complexity.

This purely function-oriented viewpoint (represented in both the function structure and the organ structure) for observing technical systems is important in discussing the nature of technical systems. It relies on the application of the concept of *organs* or *function carriers* as representing the *means* by which the functions (as aims of the TS - any effects that the technical system should exert) can be realized. It must be clear at this stage that each of the functions can be realized by a number of different organs (this factor will feature in proposition 7.4). The organ that is selected to perform a particular function in a TS determines the functions that will need to be performed by one or more of the technical system of the next lower (simpler) hierarchical level. Organs can be treated at various levels of abstraction, which in turn determines the level of abstraction needed for the evoked additional partial functions.

5.1.3 Anatomy of Technical Systems

Technical systems can only be produced if the individual constituent parts (components) can be manufactured with suitable properties of material, mass, size, form, tolerance, strength, etc., and if they can be assembled in the desired manner. The system of parts and sub-assemblies, and their configuration and (spatial or other) arrangement relative to one another, can be represented in a "component structure of the TS".

Many other properties (to be described in chapter7) must also be realized by means of the manufactured and assembled components, in some useful manner. Primarily, the functional parameters must be assured (compare chapter7) by choosing suitable configurations, arrangements, sub-assemblies and parts. Other physical and operational properties, such as the necessary load capacity and required strength, permitted deformation, corrosion resistance, life-span, reliability, ease of servicing and ease of maintenance, must each attain the desired (or at least adequate) measures. In addition, optimal transportability, minimum space requirements (for storage or use), good appearance, and beneficial manufacturing costs may be required. Each of these properties is measurable or assessable on a suitable scale of measurement. These individual measures must be achieved in a suitable combination to achieve a best compromise of all properties for the particular use and user. The means by which an engineer can achieve these other properties are referred to as the design properties, and are discussed in chapter 6 and figure 7.16.

Considering the technical system with respect to its anatomy shows a set of constructional parts (usually labelled "components") that have well defined relationships with one another in terms of their geometric, energetic and other properties. Such a viewpoint is easily understood, and is well known to most people. Almost everyone has attempted at some time or other to repair a bicycle, a water tap, an electrical connection cable, or a domestic appliance, and for that reason had to dis-assemble the object. The person has thus gained an "X-ray image" of the anatomy and component structure of the device.

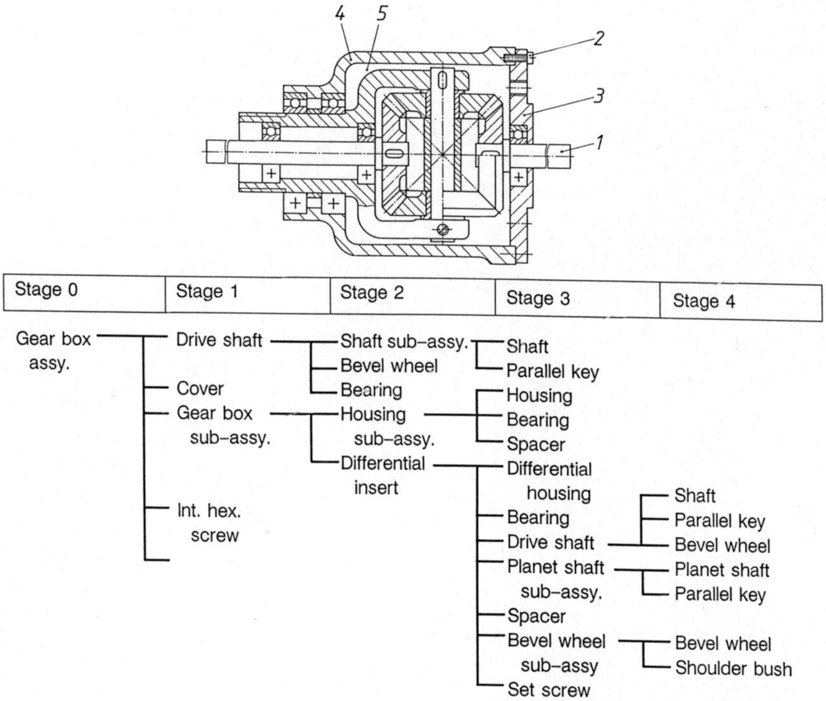

Fig. 5.1 Assembly Stages of Gear Box and Organization of Parts List to VSM 80

Modern engineering practice shows that the *component structure* of a manufactured technical systems is accurately described by the detail and assembly drawings, and the parts lists (on paper, or any other suitable information-carrying medium such as computer disk or tape), prepared according to the prevailing standards (e.g. [10]), see for example figure 5.1. The product and its description is structured according to the prescribed sequences needed to assemble it. The resulting grouping of subassemblies is guided exclusively by the demands of production, and results typically in four levels of pre-assembly of components, as shown in figure 5.1.

5.1.4 States of Technical Systems

In its progress from origination (conception and manufacture) through its working life to liquidation (to be discussed in chapter 10), each technical system attains a series of typical states of existence that occur under a characteristic set of boundary conditions, and show a different composition of the transformation system. When a particular technical system is discussed, modelled, or represented, then the corresponding state must be mentioned in order to establish the assumed viewpoint and to define the relevant transformation system.

A few of the important states are listed in figure 5.2. This figure also contains some explanations, and characterizes these states according to abstraction, and the

State of Technical System		Definition Examples	Characteristic of TS–state	
			abstract – a concrete – c	operand – Od operator – Ot
1	Planning state	When planning technical systems.	a	Od
2	Origination state	During designing, preparing manufacturing. Various stages are reached that conform to the defined structures of TS.	a	Od
3	Disassembled state	TS as sum of groups (sub–assemblies) or components; after manufacture or dis–assembly.	c	Od
4	Assembled state	TS as unit that exhibits the spatial relationships between components; after assembly, but not in operation.	c	Od
5	Testing state	TS during testing, a form of operation to determine its functional capability.	c	Od Ot
6	Storage state	TS assembled (or partially) but not installed, for longer time periods.	c	Od
7	Transport state	TS during change of location (by truck, ship, rail, air).	c	Od
8	Operation state	Working TS being used; TS exerts effects onto the operand, or is being prepared to exert effects.	c	Ot
9	Servicing state	TS during cleaning, maintenance or repair.	c	Od
10	Liquidation state	TS being scrapped or parts and materials recovered.	c	Od

Fig. 5.2 Survey of States of Technical Systems

role played by the TS within the relevant technical process (as operand or as operator).

The design engineer should mentally run the proposed system through all of these states, in order to examine how suitable the system is likely to be for the requirements in each state. Experiments or tests on models can assist in cases when the imagination is inadequate.

5.2 Model of Technical Systems – "TS Model"

The qualitative statements made in the previous section can now be formalized into a general model of a technical system. It should conform with the models of transformation systems (figure 3.1b) and technical processes (figure 4.1), and represent the nature of the technical system – see figure 5.3. As an element of the transformation system, the technical system is actively connected with the other elements of the transformation system. This infers that the TS is mutually coupled to the following:

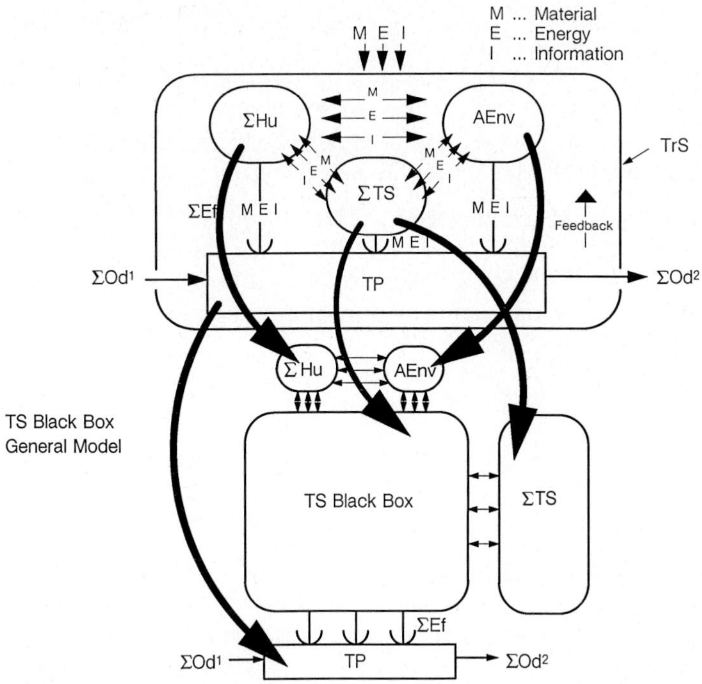

Fig. 5.3 Derivation of Model of Technical System from Transformation System

- to the operands, through effects that achieve the purpose of the TS, and cause the transformation

$$Od^1 \rightarrow Od^2;$$

- to humans as operators, for instance through the operating and regulating effects;
- to other technical systems in the transformation system, for instance when aggregating technical systems to achieve a coordinated total effect, such as regulating and controlling;
- to the active environment; for all TS a supporting connection with the "fixed system earth" has to be achieved, either directly or indirectly, and other connections made, particularly for delivering energy to a TS, e.g. from an electrical power socket.

Apart from these planned connections and couplings there exist undesired and sometimes indeterminable secondary effects, both from the environment onto the TS, usually termed *disturbances*, and from the TS onto the environment, including waste and *pollution.* All the external effects from a TS onto its environment are active couplings, and can be shown in the diagram as output effects. Every input effect on the TS from the environment evokes an output effect from the TS, according to the principle of action and reaction. Effects related to the operand and the human system usually occur only during the operational state of a technical system (see section 5.1). In contrast, the connections with other TS and the fixed system "earth" are present in the other states of the TS.

The situation described above is represented in the "Black Box" diagram of the TS, figure 5.4-A, using a simplified set of symbols.

The question now arises: How, and *with what means* can one achieve those planned output effects on the operand, which are the real tasks that have been established for technical systems? Interpreting the explanations of section 5.1, a number of answers are possible. The effects can be achieved if:

(1.) certain well-defined effects (coming from the human, other TS and the active environment) exist that form the inputs to the TS under consideration, AND
(2a.) certain well-defined tasks (functions) are fulfilled by the TS, in which each of these function depends on one or more relationships of input to output and on the selected mode of action, OR
(2b.) certain well-defined functions and their relationships as mentioned in (2a.) exist, together with the organs and their relationships that fulfill the functions, OR
(2c.) certain well-defined organs and their relationships as mentioned in (2b.) exist, together with the realizable constructional elements and their relationships.

It may be recognized that the alternate answers (2a.), (2b.) and (2c.) describe the interior of the TS at various levels of abstraction, namely:

(2a.) at the level of the function structure, figure 5.4-B, a structured arrangement of TS functions and their relationships;
(2b.) at the level of the organ structure, figure 5.4-C, a structured arrangement of TS organs and their relationships;
(2c.) at the level of the component structure, figure 5.4-D, a structured arrangement of components and their relationships.

At this stage of our discussion of the theory of technical systems these terms are not precisely defined, but can be better understood by considering the example of a machine vice shown in the right-hand column of figure 5.4. The individual levels of abstraction and the relevant structures are demonstrated on this simple machine system, and correlated with the TS model. Some of these models use symbolism which should become clear from the figure and the following discussion.

A diagram of one of these structures of a technical system is necessarily an abstraction from the reality. Abstractions can be produced at various levels. During any process of abstracting, some information about specifics may be lost, but other information about context must be added.

The abstraction of an existing technical system that is most easily produced is an assembly drawing. It contains information about the form and shape of each part of a specific artifact. Consider, for instance, the assembly drawing of the machine vice shown in figure 5.4-D, which represents the component structure in its most concrete form. Additional detail is shown in the detail drawings for each component, parts lists, manufacturing planning documents, CAM (Computer-Aided Manufacturing) and NC (Numerical Control of machine tools) programs, etc.

Abstracting can take place in two directions shown in figure 5.4. One of these is related to generalizing the concept of that type of structure, the other leads to dis-

GENERAL MODELS – IN OPERATIONAL STATE	*EXAMPLE – MACHINE VICE IN OPERATIONAL STATE*	
	ABSTRACT	*CONCRETE*
LEVEL 0 Design Specification		
Requirements Specification (from customer, potential user)	Contract Specification (company level)	Designer's Working Specification (designer's approved interpretation)

LEVEL A TS – Black Box

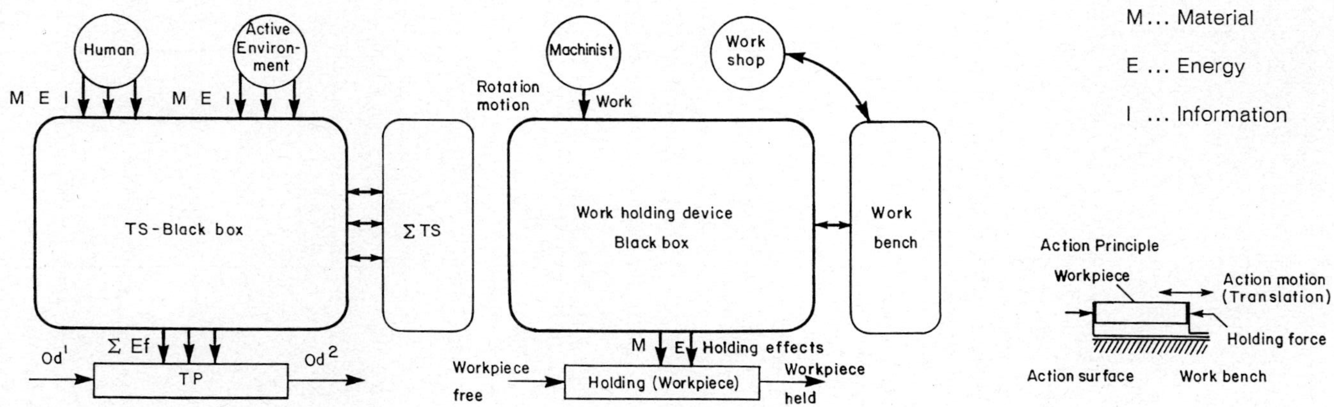

Fig. 5.4 "TS Model" – Models of Technical Systems, General and Example: Machine Vice

ABSTRACT *CONCRETE*

LEVEL B Function Structure

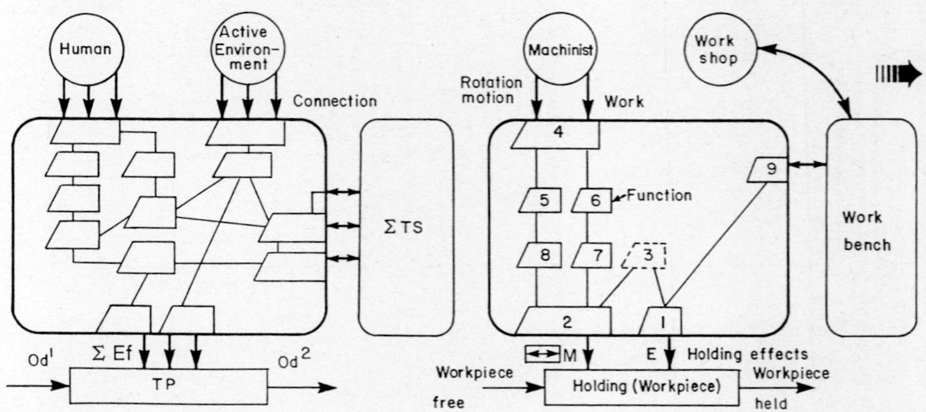

	Function	Organ	
1	Fixed Holding Surface FORM Holding Action TRANSMIT	Fixed Jaw	FB
2	Movable Holding Surface FORM Holding Action TRANSMIT (Holding force)	Movable Jaw	BB
3	Movable Holding Surface GUIDE relative to Fixed Surface Relationship between Functions 1 and 2	Guidance BB ↔ FB Moments, Radial Forces	FB
4	Action Surface FORM Work(force X distance) ACCEPT	Acceptance - Organ	UO
5	Rotation to Translation Movement TRANSFORM	Motion Screw Organ	SM
6	Force AMPLIFY		
7	Energy for Holding STORE (strain energy)	Storage Organ	SO
8	Position RETAIN (against Holding Force)	Motion Screw Organ SM ↔ FB SM ↔ BB Axial Forces	SM
9	Connection Surface to Fixed System FORM Connection PERMIT	Connection Organ	VO
		Motion Screw Organ SM ↔ FB Radial Forces	SL

Fig. 5.4 (cont.)

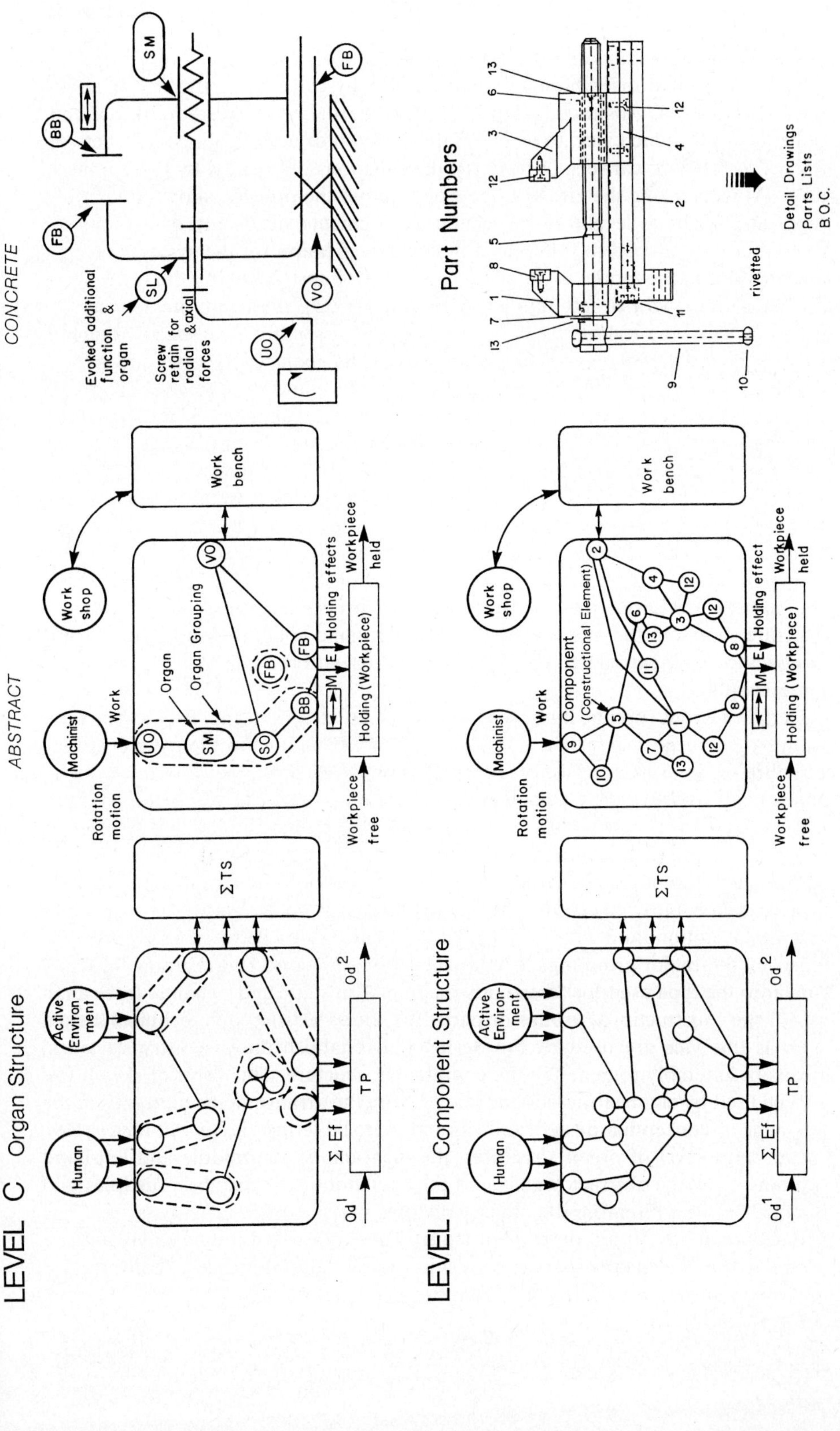

LEVEL C Organ Structure
LEVEL D Component Structure
ABSTRACT
CONCRETE
Human
Active Environ-ment
Σ Ef
TP
ΣTS
Machinist
Work
Rotation motion
Organ
Organ Grouping
Holding effects
Holding (Workpiece)
Workpiece free
Workpiece held
Work shop
Work bench
Evoked additional function & organ
Screw retain for radial & axial forces
Component (Constructional Element)
Holding effect
Part Numbers
rivetted
Detail Drawings
Parts Lists
B.O.C.

important meaning and orientation in these models. The terminology can be related as follows:

- the effects of the TS (as aims) are achieved by certain function structures (as means);
- these function structures (as aims) can be realized by various organ structures (as means);
- the organ structures (as aims) can only be realized from various component structures (as means).

These TS structures are discussed in more detail in the following sections.

5.3 TS Function Structure

The TS as the object being observed is defined for the purpose of this structure in terms of its functions. In this way the function structure is generated or derived as a collection and interrelationship of functions. It is defined by a set of its elements (functions) and a set of relationships of these functions to one another. The function structure gives the engineer a means to evaluate the operational states of a TS.

5.3.1 Function

The function is a property of the technical system, and describes its ability to fulfill a purpose, namely to convert an input measure into a required output measure under precisely given conditions. By using the word "set" in the third sentence of the opening paragraph of section 5.3, the TS function can be understood as a unique coupling of the elements of a set of *independent* input measures to the elements of a set of *dependent* output measures. This interpretation shows some similarity to the concept of a mathematical **transfer function** as defined for dynamic systems.

As an example, the screw spindle of the machine vice in figure 5.4 can be held in bearings at a certain load and rotational speed. Alternatively we could speak of positioning the spindle by limiting it to one degree of freedom. A spindle with 6 degrees of freedom is brought into the state of one degree of freedom (rotation) by the function "holding in bearings".

5.3.2 Relationships Among Functions

These are mainly couplings between individual functions. The function structure of the machine vice shown in figure 5.4-B can be interpreted as follows: the energy of rotational movement from the hand is accepted (function 4), the force amplified (6), rotation converted into translation (5), and translation with force is transferred to the jaw (2). In this way a plausible representation of the mode of action is generated.

5.3.3 Types of Functions

Many different kinds and classifications of functions appear in the literature. Here we will only define a few basically important types which will be used in the further discussions. Of the many classifying viewpoints, three are particularly meaningful:

(a) **Complexity of the function.** Each function may be assigned to a certain degree of complexity in a hierarchy of complexities. The lowest degree is occupied by the elementary functions, those that cannot (usefully) be resolved into more limited functions.

(b) **Degree of abstraction of the function.** Each function may be described at various levels between "concrete" and "abstract". This in turn influences the number of possible or available organs (function carriers) that can be found (as means) to realize the function (as the aim). If, for instance, the given description refers to the function "alter motion", then the range of available means is very broad. With an increased number of additional data about the function, selected from the ranges of effects, conditions, operations or working means, the range of available means of fulfilling a function is progressively narrowed, until a single concrete TS remains. The additional data mentioned here are the design properties, compare figure 5.12.

The degree of abstraction is derived from the design properties for the given ranges. Figure 5.5 shows a general scheme for this hierarchy of abstractions (an example showing the function "transporting" may be found in figure 7.2). The

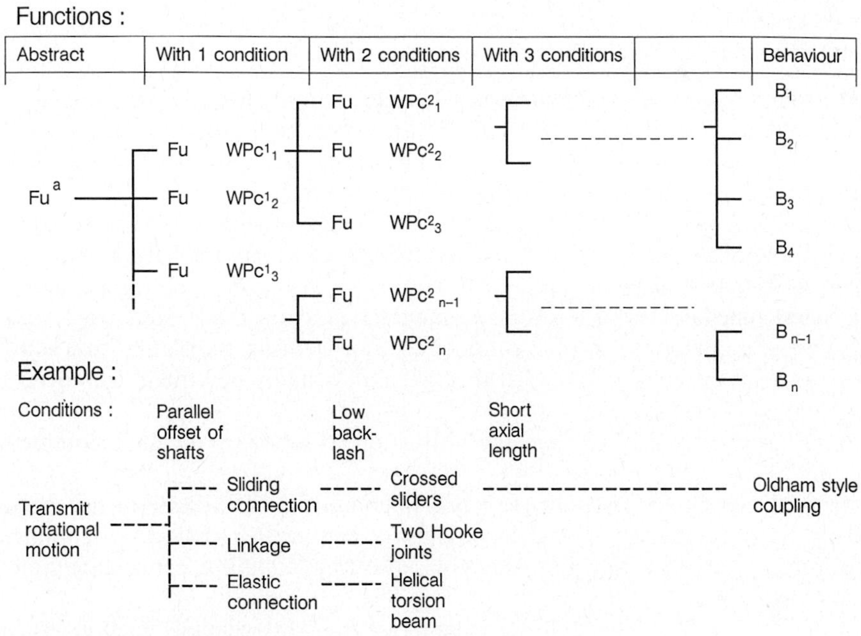

Fig. 5.5 Degrees of Abstraction of Functions in Technical Systems

general scheme shows the functions designated as "functions with **i** conditions". The order in which these conditions appear is not determinate, the behavior of a real TS can also be attained by a different order. For systematic work in engineering design it is essential that a certain ordering arrangement is selected and agreed, and used consistently.

The connection between the degree of abstraction and the degree of complexity should also be noted. Resolving the functions into partial functions (i.e. functions of lower degree of complexity) is only possible and sensible when a certain more concrete level of abstraction has been obtained. In our recommended procedural scheme this is true only when an action principle has been established. Designers who are familiar with systematic design methods will recognize an analogy to the morphological chart [119, 214], in which the path of progress leads from a function, by considering one or more action principles, to the organs (function carriers) able to fulfill that function. Others of these characteristics are unconsciously neglected, either because they are implicit in the statement of the task, or they are assumed by tradition to be "fixed variables".

(c) **Categories of purpose of the functions**. The transformation function, with the selected mode of action, that fulfills the purpose of the TS is necessarily accompanied by a series of additional functions. These are essential, and serve to ensure that the transformation function can be realized, or its operation supported. Based on statement 4.2 (see section 4.12), these additional functions are:

- auxiliary functions (with respect to the transformation function, but not with respect to their importance),
- driving, propelling or energy delivering functions,
- regulating and controlling functions, with appropriate sub-functions such as "measuring",
- connecting and supporting functions that make the TS into a "unit".

Apart from the three types of functions discussed in sub-sections (a), (b) and (c), a few other terms that frequently occur in the literature should be defined. They are not a part of the theory of technical systems as presented in this book, but show a connection to other theories and methodologies of engineering design.

(d) **Logical function**. This type transforms one or more independent variables into a single dependent variable that can only take on two measures (e.g. 1 and 0, + and −, or true and false).

(e) **General function** [187]. This is an elementary function that results from coupling a general operation (storing, conducting, transforming, translating, and summative or distributive associating) with a general "value type" (material, energy, information).

(f) **Basic function** [182] is an elementary function (combining, dividing, conducting).

(g) **Physical elementary function** (or basic operation) [143]. This term has been applied to 12 elementary functions: emitting, conducting, collecting, guiding, transforming, enlarging, direction-changing, directing, coupling, connecting, adding and storing.

The functions defined in (e) to (f) serve to establish the general function structure that contains the elementary functions standardized for that particular

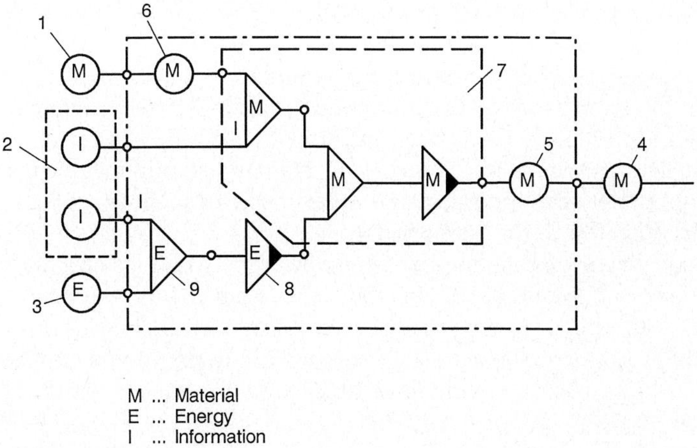

Fig. 5.6 General Function Structure; Example: Coffee Grinder [186]

method. These elementary functions are based on the idea that only the three flows of material, energy and information can be present in technical systems, and that only certain operations (basic operations) are permitted. An example of such a general function structure is shown in figure 5.6.

Note that the term "total function" is not included in this list, it is more related to an organ structure, see section 5.4.

5.3.4 Resolution of Functions

In a hierarchy of functions, the ways in which a higher function of a technical system can be resolved or decomposed into its partial functions should now be investigated. This resolution is not as easy or obvious as examples quoted in the literature might suggest. The classes of partial functions into which higher functions can be resolved are of three types:

(a) partial functions that are determined from, and because of, the selected mode of action;
(b) partial functions that are necessarily evoked by the given function (see sub-section 5.3.3, item (c) categories of purpose of functions);
(c) partial functions that serve to realize other necessary properties of a TS - manufacturability, transportability, safety of humans, etc. (i.e. additional tasks) - such as a "connection between sub-assemblies" made necessary by reasons of final erection or transportation, or a "guarding" of moving parts needed to protect humans from danger.

Classes (a) and (b) are established while working out the function structure and/or organ structure. Class (c) is particularly useful during the transition from the organ structure to the component structure.

5.3.5 Representation of Function Structures

Figure 5.4-B shows a representation of the function structure that conforms to the general arrangement of elements of the transformation system. A typical arrangement of the functions within such a function structure should reflect the general model of function structures (see figures 4 and 20 in [119]) and should include the auxiliary, propelling, regulating, supporting and other evoked functions, and the receptor and effector functions at the TS boundary.

This representation is by no means the only possible form. Many examples of other forms are described in the literature [14, 167, 187]. Especially noteworthy is the form used for the "general function structure" according to Roth [187]. It uses the "general functions" as defined above (sub-section 5.3.3, item (e)) and can be regarded as a form of logic flow or switching schematic. An example is shown in figure 5.6.

Another form that is similar to the function structure is that of the function tree [14]. It shows the functions with only their immediate dependent relationships, but cannot easily show all other relationships, especially between functions in different branches of the tree structure. Nevertheless, because it is a clear and simple representation, it finds frequent use. Examples are given in the literature, for instance figure 21 in [119].

5.3.6 Establishing Function Structures and Their Assertions

There are two main ways to work out a function structure. One starts by abstracting from the component or organ structure, and is *analytical.* It results in either a very comprehensive function structure with many types of function (compare sub-section 5.3.4, resolution or decomposition), or, by neglecting those functions that are not essential for the direct transformation process, a function structure similar to the one established from the "black box" can be obtained.

The second way starts by concretizing from a "TS black box", and implies that the functions and their relationships are to be found or *synthesized,* as the means by which the output effects that will drive the technical process (as the aims) can be realized. This path leads through establishing the inputs and modes of action of the TS, and can take place at various levels of abstraction and achieve differing levels of completeness, as explained above.

Usually, even at the beginning of a project, a number of the characteristics and properties of the technical system are already firmly established and fixed, typically:

- the effects (as output from the TS) needed to drive an established technology of transformation;
- the mode of action, defined by action sites, action motions and conditions, see figure 5.7;
- the degree of mechanization or automation in executing the technical process;
- even at times the actual basic mode of action of the TS and its inputs.

The number of these design characteristics that are established in this stage of engineering design, when considering the function structure, depends on the

Character of Effect	1st Stage Characteristic		2nd Stage Characteristic	
		State of Embodiment		State of Embodiment
Effect	Type	Action motion	Type	Degrees of freedom – combinations,
			Sense	(see figure 5.10)
			Measure	→← ←→
				Velocity, Acceleration
			Timing	Duration, frequency
			Number	1 ...... n
		Action force	Measure	Force unit
			Sense	→← ←→
			Timing	F = f(t)
			Number	1 ...... n
		Action heat	Measure	Thermal energy, temperature
		Action field	Type	Electrical, magnetic
Action Site	Genus	Space, surface, line, point		
	Type	Direct –– solid Indirect –– action medium		
	Form	Geometric bodies		
	Position	Horizontal, vertical, ...		
	Measure	Length unit		
	Number	1 ...... n		
	Surface	Hardness, Roughness ...		
Action Conditions	Medium	Gas, liquid		
	Pressure	Pressure unit		
	Temperature	Temperature unit		
	Field	Electrical, magnetic		

Fig. 5.7 Effects of Technical Systems, Descriptive Characteristics and States of Embodiment

degree of abstraction, and determines the degree of completeness of description for the TS (compare figure 5.11).

5.4 TS Organ Structure

The organ structure is an abstract model of a technical system that includes the organs (the carriers of the functions, as the means) that realize a certain class of modes of action (as the aims), and the relationships between those means. Organs may be collected into useful groups (organisms) that together perform a more complex function. Questions about the material and physical embodiment are not considered. The organ structure permits analysis and representation of the operational states of a technical system. The organ structure must contain at least one organism of the action chain, and an organism for connecting to the fixed system (the external world, usually our planet earth).

5.4.1 Organ

An organ is defined as a system that realizes a given internal function of a technical system. Organs may also be termed "function carrier" (also used in this book) or "functional unit" (used in the general literature, but this term is usually more ambiguous, it can also refer to a constructional unit or sub-assembly). The term organ is used in biology in the same sense and with the same content as in this theory of technical systems. In the physical sciences it is also not a new concept, Reuleaux [179] and Leyer [150] have used it, but not as strictly defined as in this book.

As with the functions (section 5.3) and the constructional elements (section 5.5), the organs can be defined at various levels of abstraction. At higher degrees of abstraction, an organ can be designated by a generic term for a group of function carriers (e.g. "bearing") or shown as a symbolic representation (e.g. figure 5.4-D). At lower degrees of abstraction the designation can be more specific (e.g. bearing with angular force transmission and rolling contact) and the representation can be more concrete (e.g. the evoked organ "SL" in the concept sketch of figure 5.4-D). The latter example is still symbolic, but it delivers a visual image that is quite close to the material reality. Note that in this context we are concerned with the *organ* that transmits force and permits rotation (its functions), and this includes the relevant surfaces of the shaft and the housing. The sub-assembly "angular contact ball bearing" which consists of the two ball races, the rolling elements and the cage is a component (section 5.5), and realizes a part of the organ.

5.4.2 Relationships Between Organs

The main relationships between organs are couplings: the output from one organ is the input to the next organ in the action chain. These couplings can take on a number of different forms. The relationships are already present in the abstract organ structure. In the representation of the mode of action they are shown as the connections between organs. Representing the organ structure in a *concept sketch* (or sketch of principles) defines the spatial relationships between organs, at least in a rough and approximate way.

5.4.3 Types of Organ

Organs and functions are conceptually very similar, and consequently their classifications (and classifying criteria) are also similar:

(a) complexity of organs – a (relative) ranking can be found, such as: organisms, groups of organs, organs, and partial organs;
(b) degree of abstraction of organs – as with the function structure, organs can be relatively more concrete or more abstract;
(c) categories of purpose for organs – these are the same as for the functions: transforming, auxiliary, energy delivering (propelling), regulating or connecting organs (see also the next sub-section).
(d) location of organs with respect to the boundary of a TS:
   - organs at the boundary, either receptors or effectors,
   - internal organs.

The effector organ at the boundary that realizes the output effects of the TS (and delivers them to the operands of the process) is called the "transformation organ" (see item (c) above).

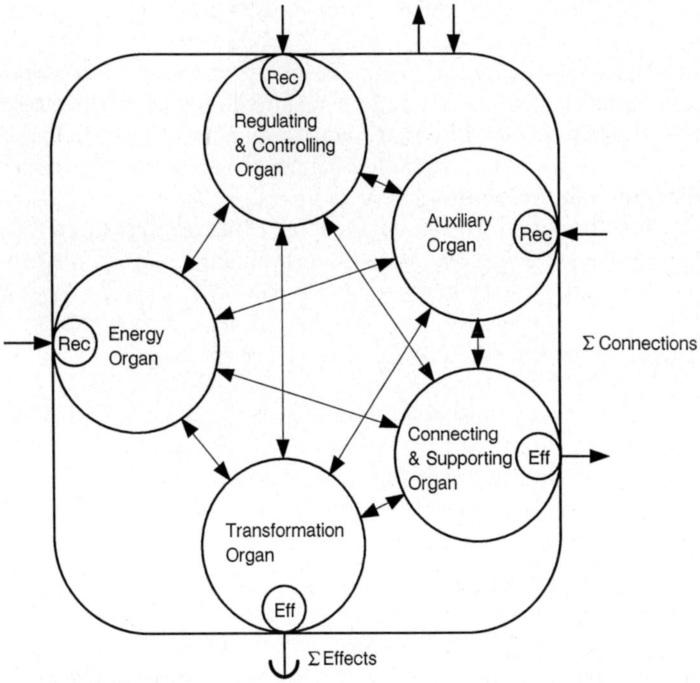

Fig. 5.8 Transformation Organism Comprised of All Types of Organ

5.4.4 Transformation Organism - Units

Individual organs within the organ structure, when suitably linked to the transformation organ, form an organism - a transformation organism - which is embodied as a characteristic unit within the topology of an organ structure. Figure 5.8 shows such a transformation organism, built up in a similar way to the "complete function" or "total function" as used in the literature [14, 167].

5.4.5 Types of Organ Structure

Mechanical engineers may find some difficulty in grasping the concept of an organ structure. They are more used to decomposing a machine system into constructional units (sub-assemblies and components) such as screw and nut, gear pair with shafts and bearings, etc. Organs have different purpose, they provide the functional connections between parts (components).

Other engineering disciplines have worked with the concepts of organs for some time. Such organs may be called switches, condensers, relays, resistances, etc. in electrical engineering; or pressure cylinders, valves, etc. in hydraulics and pneumatics. They are generally assembled into organ structures in the form of "circuit schematics", showing only the organs and their connections. The wiring or piping diagrams represent the appropriate component structures, and indicate the physical and geometric arrangement in the available space. The situation is simpler in these fields because the major organs usually correspond to the actual components.

Organ structures are usually generated when designing on projects of large-scale plant. The structural element is generally a technical system of the third degree of complexity, usually termed "machine" and "apparatus". Some fields of engineering have developed a special symbolic language for representing organ structures (e.g. hydraulics and pneumatics).

A similar example may be found in civil engineering. Structuring a space into rooms, as done by an architect, is nothing more than formulating an organ structure.

5.4.6 Representation of Organ Structures

Organs are object systems, and can be graphically represented by circular symbols, according to our conventions.

As mentioned in section 5.3 and shown in figure 5.4-C, many different representations of organ structures exist, and usually employ a specialized and standardized language and symbolism. Examples are:

- semi-conductor elements (DIN 40700)
- electro-technical circuit schematics (DIN 40710, SEV 9001)
- hydraulic schematics (DIN 24300, VSM 10332)
- wiring diagrams (DIN 40717)
- piping and pipe fittings (VSM 10330), see also figure 5.9
- thermal power plant (DIN 2481).

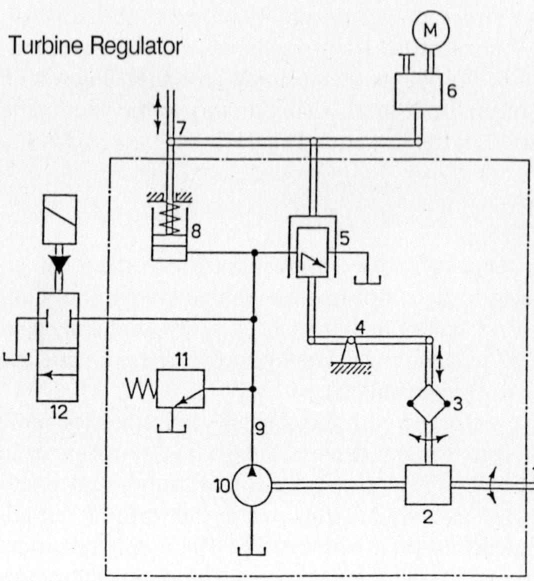

Symbols according to VSM 10332

Designation

1 Turbine shaft
2 Gearing
3 Centrifugal regulator
4 Fixed fulcrum (pivot)
5 Sensor valve
6 Adjustment drive, motor actuated with hand adjustment
7 Regulation linkage
8 Servo-motor (cylinder)
9 Pressure piping
10 Pump
11 Pressure limiting relief valve
12 Shut-off valve with solenoid-driven return force

Fig. 5.9 Organ Structure as "Switch Schematic": Example

At present no such uniform symbolic representation is accepted in mechanical engineering for use in concept sketches, although a few recommended sets of symbols are in existence.

5.4.7 Establishing Organ Structures and Their Assertions

As with the function structure, it is possible to establish an organ structure either by abstracting from the component structure, or by concretizing from the function structure of a technical system.

Starting from the function structure, a few of the design characteristics must be established, in part or fully. Examples of these design characteristics, and how completely they must be established, are:

- the inputs and outputs of a TS at its interaction points with other systems (these design characteristics need to be fully established);
- the mode of action of the TS and the organs capable of fulfilling the functions, aggregation of tasks for organs, and rough, approximate spatial arrangement.

The number of items (organs or design characteristics) that are established by defining the organ structure, and the amount of information they provide, depends on the degree of abstraction, and determines the degree of completeness of description for the TS (compare section 5.6).

5.5 Component Structure of Technical Systems – TS Component Structure

The component structure represents the most concrete way of representing a technical system, and the most concrete stage in generating the TS during the design process. This structure must therefore contain all the necessary properties that were given as requirements, and that need to be established by designing to fulfill those requirements. The number of criteria now needed for evaluating the technical system (based on its component structure) must increase by a large amount, compared to the criteria used for evaluating the more abstract structures, which were selected and assessed mainly from a functional viewpoint (for further discussion of evaluation see chapter 8).

The component structure is a description of a technical system which refers directly to manufactured (or manufacturable) constructional elements. It thus contains the set of such elements and the set of their relationships. This description may show either an operational state of the technical system (for instance a design lay-out contains the connected systems and their environment), or the "disassembled" or "assembled" states (e.g. detail and assembly drawings).

5.5.1 Constructional Element

The component structure is composed of *constructional elements* (components) and the relationships between them. Such a constructional element may be defined in terms of any of the levels of complexity (as will be discussed in chapter 6). The definition depends on the complexity of the system being considered, and the origin of that element. A constructional element may originate by manufacture within the organization (in-house, self-manufactured) or by acquiring it from another supplier (out-sourced, bought out, for OEM – "original equipment manufacturers").

The term "constructional element" is not an absolute designation, it depends on the context. A complex building or a complex transfer machine line may be regarded as a constructional element for a plant (TS of high degree of complexity), and a simple pin may be appropriately termed a constructional element for a jig or fixture. From the viewpoint of the manufacturing organization, the term is even more relative, which can be seen by comparing groupings in the parts list from figure 5.1, showing four levels of parts or sub-assemblies used as constructional elements for the next higher level.

Usually, the description to be used for a standardized constructional element is prescribed in a document (national or international standard, code of practice, etc.). The standard code defined for a metric machine screw is M 10 × 40 DIN 912. A different form of a brief description of a component may be found in a manufacturer's or supplier's catalog, and uses the specified trade name and ordering code (catalog number).

The manufactured constructional elements (components, machine parts, codes for transmitting information) are defined by means of the applicable design properties – form, material, dimensions, tolerances, surface finish, etc. (refer to chapter 7 for further discussion).

5.5.2 Relationships Between Constructional Elements

If the discussion is restricted to the assembled state of the TS, conforming in the first instance to the output state of the design phase, and secondly to the output of manufacturing, then three types of relationship may be recognized in the component structure:

- spatial relationships, achieved by locating the elements in space;
- mechanical coupling between the individual elements, described by the degrees of freedom (see figure 5.10);
- energy relationships, mostly stating the forces between parts (e.g. the statement of torque or force to pre-tension the bolts in a connecting rod of a reciprocating piston machine).

Some of the energy and other relationships are a consequence of the constructional elements and their spatial arrangements, for instance the relationships "sealed airtight", "electrically isolated", "thermally insulated", or the spatial arrangement "shrink fit".

5.5.3 Types of Constructional Elements

Various types of constructional elements can be identified by considering the following criteria:

Nr.	Degrees of Freedom: Translation **T**	Rotation **R**	Σ	Examples of Known Mechanical Couplings	Symbolic Representation
1	0	0	0	Fixed connection	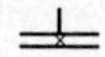
2	1	0	1	Linear guidance	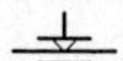
3	0	1	1	Rotational connection	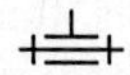
4	2	0	2		
5	0	2	2	Hooke joint	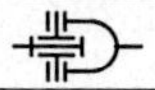
6	3	0	3		
7	0	3	3	Spherical joint	
8	1/2	1/2	1	Screw pair	
9	1	1	2	Rotation & translation	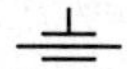
10	1	1	2		
11	2	1	3		
12	1	2	3		
13	2	1	3	Flat pair	
14	1	2	3		
15	2	2	4		
16	2	2	4		
17	3	1	4		
18	1	3	4		
19	3	2	5		
20	2	3	5		
21	3	3	6	No coupling	

Fig. 5.10 Mechanical Couplings Using Differing Degrees of Freedom

(a) Complexity of the constructional element: part, TS group (at different levels - sub-assembly, assembly, etc.), TS machine, apparatus, works of building construction, plant, equipment (at different levels).

(b) Degree of abstraction: ranging from an abstract designation that covers a class of constructional elements, to the definition of a single element described by its design properties (compare figure 5.4 and 5.11).

(c) Purpose of the constructional elements: this covers a wide range of types, specific for each branch of engineering. In mechanical engineering, the designation of constructional elements is closely associated with the names of common machine elements.

(d) Constructional elements and their location relative to the boundary of the technical system:
- internal constructional elements;
- border elements, that provide the links between internal partial systems (e.g. between sub-units to be assembled "on site" during final erection of the TS);
- boundary elements, that are present at the external boundary of the technical system in its state of "finality" and directly exert the effects on the operands; they embody the effector locations, and are capable of performing the action motions.

5.5.4 Mode of Construction

The component structure depends not only on the effects to be exerted, the mode of action, and other properties that must be realized by the system. It also depends on how the principle and policy which helps in managing the processes of constructing (building and manufacturing) a technical system is interpreted. This principle is termed the *mode of construction.*

Very typical of modern equipment is the mode of construction known as modular. It can be found in many types of technical systems, and has advantages both from the point of view of manufacturing, and of operating the system.

Frequently, a combination of such policy directions (and their corresponding modes of construction) help to decide how the construction of a technical system will be accomplished. Some constructional elements may be modular, some may comprise only components purchased from outside suppliers, some may be subcontracted, some may be interchangeable only as sub-assemblies, etc.

5.5.5 Representation of Component Structures

The description of the component structure in most cases contains the prescriptions for manufacture and assembly. Compared to the more abstract structures, the methods and tools for representing the component structure have long been customary, and standards have been issued to regulate the preparation of the necessary documents. These standards and customs have evolved over the centuries [38, 45] as a result of improving manufacturing technologies, of working means available to designers (drawing equipment, computer aids, etc.), needs for communication of manufacturing information, and international coordination.

A different situation exists in executing preliminary lay-outs (usually in sketches or rough drawings) and dimensional lay-outs (to scale, as basis for producing detail drawings) used for designing a TS. In many cases the representation may closely follow the standards for detail drawings, so that the transition from lay-out to detail is easier. Lay-outs are nevertheless different in various aspects, and their representation depends on many factors, including particular traditions established in various companies and countries.

5.5.6 Establishing Component Structures and Their Assertions

A component structure can be established by concretizing from a function structure or an organ structure, by means of a series of preliminary and dimensional lay-outs. This task is complicated because of the vast number of relationships that exist between the design requirements and the design properties of the system. Consequently, the designer's task of establishing lay-outs (and component structures) is usually an iterative procedure.

A typical sequence consists of three steps - preliminary layout (roughly to scale), dimensional lay-out (to scale), and detail and assembly drawings. These represent the component structure with increasing concreteness and completeness. The relationship of these design documents with respect to concreteness and completeness is indicated in figure 5.11.

The complete component structure defines all the design properties that have been established by considering all requirements in a step by step transition from organ structure to component structure.

5.6 Comparison of TS Structures and Their Transformations

The three TS structures described in the previous sections may be compared with respect to two significant variables. These are the degree of maturity of the TS model, and the completeness with which it describes the system. Figure 5.11 shows such a comparison with maturity as the ordinate, and completeness as the abscissa. As the description of a TS becomes more abstract, so the completeness of that description must reduce. At each level of abstraction or completeness, by establishing appropriate design characteristics and properties a greater completeness and maturity of description can be achieved.

A typical progression for designing a technical system is indicated by the diagrams and explanations in figure 5.11. The general models at each level of completeness are valid for all technical systems. Some particular technical systems and their conventional solutions are contained in the available "masters" (sometimes referred to as "paradigm solutions"). Within each modelling range (compare figure 5.4), the model of a TS to be designed increases in maturity, until it is ready to transfer to the next, more concrete modelling range (and be transformed into the next TS structure). Various other models may be developed as aids to designing and evaluating the proposals.

The result of a completed design process is typically a set of detail and assembly drawings, parts lists, etc., the most concrete level shown in this figure. Such a complete description of the technical system can only be achieved by fully specifying all the design properties of the technical system. Therefore the complexity of a technical system must also be considered. Only if a complex system has been completely resolved into all its constructional elements can the design properties be used to describe the system.

Figure 5.11 indicates only a qualitative and rough guide to the relationships. A better guide is not easy to achieve because the scales for concreteness/abstractness, completeness and maturity are not precisely defined, and would be difficult to define even for a specific technical system.

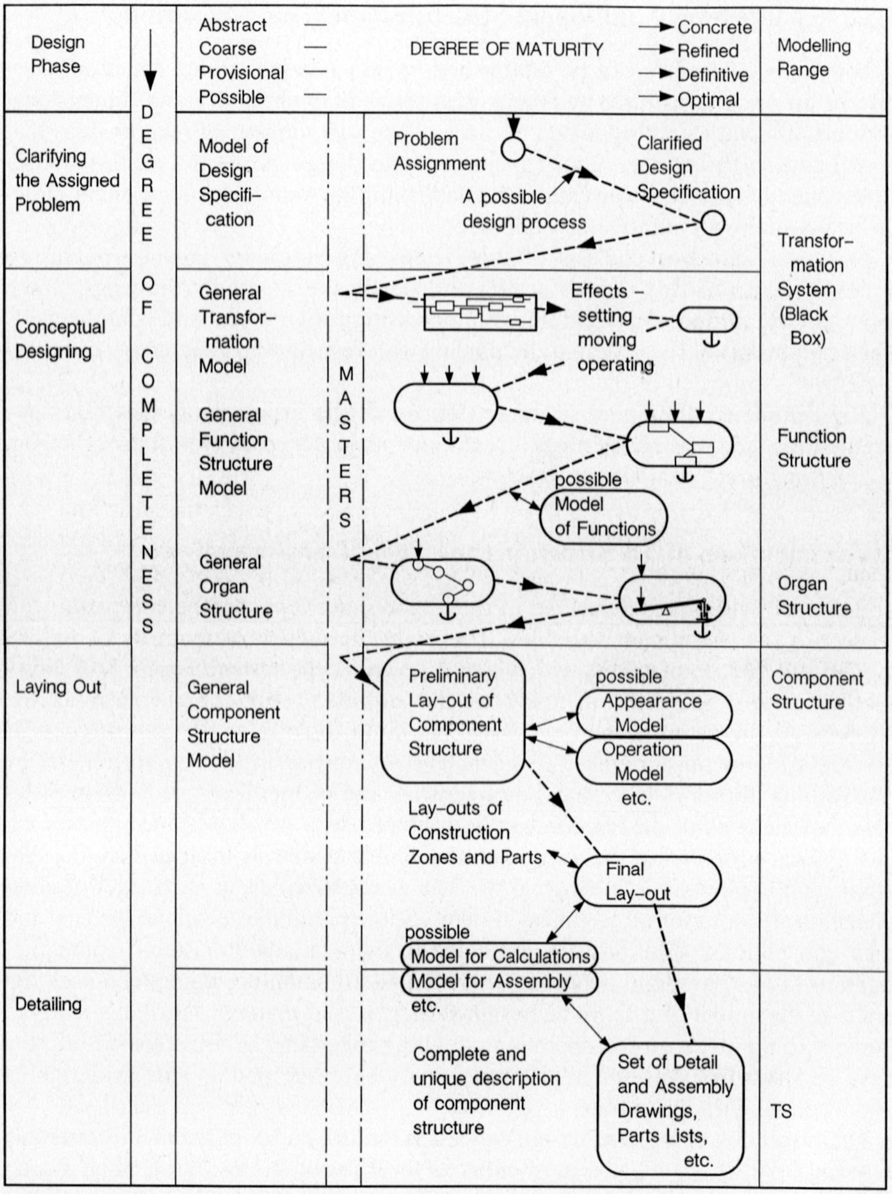

Fig. 5.11 Completeness and Maturity of TS during Design Process [127]

The transformation of a TS model from one type of TS structure (and modelling range) to another can take place in two directions:

(a) designing is achieved by concretizing, which requires the designer to establish some more of the design characteristics and properties, or
(b) experience can be generalized by abstracting from the more concrete models.

Each of the design characteristics exhibits a number of states of embodiment, because the result of each step of concretizing (transforming from a more abstract to a more concrete TS structure) offers a number of variants. Transforming to a more concrete TS structure is a one-to-many mapping where the number of available alternatives is to some extent unpredictable. A function in the function structure of a TS may specify "allow pure rotation, and transmit axial and radial force", and suggests the organs of "bearing" or "hinge" or "crossed spring joint" as its more obvious states of embodiment in the organ structure. The organ "bearing" can in turn have many states of embodiment in the component structure, such as "angular contact caged rolling element bearing", "sliding oil lubricated bearing", "sliding air pressurized bearing" (each with the appropriate support surfaces on housing and shaft), and many others. Within each of these embodiments at the preliminary lay-out level, a further set of states of embodiment concern the forms and dimensions of the constructional element, for instance "deep groove ball bearing", "angular contact ball bearing", "taper roller bearing", etc., each of which require adequate forms of housing and shaft.

The TS structures described in sections 5.3, 5.4 and 5.5 (and others that may be found) are in a homomorphic relationship with one another. It is possible to reach a single conclusion when moving in one direction from a component or its relationships in one TS structure to a component and its relationships in a second (usually more abstract) structure, but not in the reverse direction.

Transforming the structures of technical systems comprises the content of many engineering tasks. The process of designing is particularly concerned with transformations from more abstract structures to more concrete ones - concretization. This theory should be particularly important for systematic and methodical design procedures, because it forms a necessary basis for guiding iterative progress, and formulating an algorithm for engineering design. Such an algorithm may in some rare and specialized cases be complete, and therefore can be transferred to a computer. In most cases the available algorithm will be incomplete for the design task, and will require interactive human intervention to provide the creative and imaginative components (see also the further discussions of design in chapter 12).

The design characteristics that are established for each of the ranges of application for the TS structures are listed in figure 5.12.

5.6.1 Relationship Aims - Means

We have already discussed this relationship that is so important for design engineering; a few of these relationships are now explained in greater detail. Especially the references to "finality" are important in the treatment of causality (section 5.1).

The relationship *aims - means* expresses the "link towards finality", and also represents the process of synthesis by which a designer, starting from the aims, searches for appropriate means to fulfill the system of goals (i.e. the requirements and needs).

When considering complex systems of goals ("why?"), the question about appropriate means ("with what can the aims be achieved?") must be asked a number of times before a level of elementary means (that are completely describable)

Region	Design Characteristic		Represented by	Comments
	Characterization	Examples		
Effects Region	Characteristics which explain and establish the output effects of the TS (as a whole) in addition to the effects and action conditions.	Technological principle Technology Sequence of partial tranformations Degree of application of TS in the TP Number of TS in the TP Effects Action locations Action conditions	Technical process, see fig. 4.1	Definition from output effects, see fig. 5.7
Functions Region	Characteristics which explain and establish the internal action chains (internal tasks) through which the effects are achieved from the selected inputs.	Inputs, main secondary Action chains, mode of action Evoked function Degree of mechanization Degree of automation Aggregation of functions Decomposition of functions	Function structure, see fig. 5.4–B	
Organs Region	Characteristics which explain and establish the types (classes) of means (internal tasks) through which the effects are achieved from the selected inputs.	Action principle Evoked functions and organs Aggregation of organs Decomposition of organs ... Rough arrangement in space Connection (interaction) locations to other TS	Organ structure, see fig. 5.4–C	
Component Region	Characteristics which explain and establish the types (classes) of means and their grouping that have been selected to execute (realize) the function chains.	Mode of construction Evoked functions Strength characteristics Stiffness and elasticity characteristics Wear characteristics Hardness Corrosion resistance Heat resistance ... Principle of form determination Families of components	Component structure, see fig. 5.4–D	
– All design characteristics are realized by the design properties – Design properties also appear as requirements				

Fig. 5.12 Survey of Design Characteristics of Technical Systems

is reached. This procedure can either be described generally as a series of aims-to-means steps (whereby the means of the previous step become the aims of the next more concrete step), or in the realm of technical systems the process starts directly from the effects to be exerted (as aims) to specify the required technical systems (as means). Figure 5.13 shows such an aims-to-means procedure and interprets the

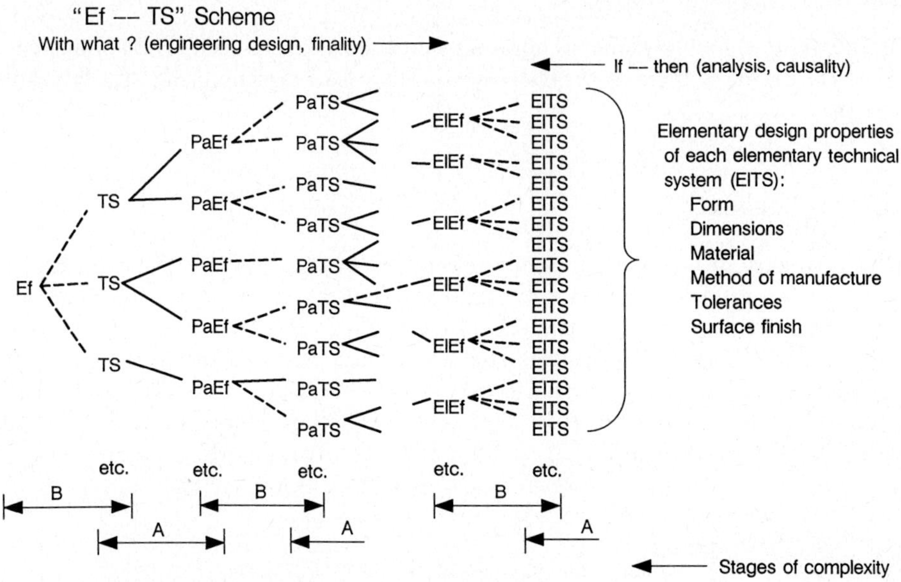

Stages "A": In order to achieve its total effects (as aims), every TS needs a sequence of certain partial effects (as means);
Stages "B": Each effect can be achieved with the aid of various TS.

Fig. 5.13 Scheme of "Goals - Means" or "Effects - Technical Systems"

individual steps of "why - and with what?". This scheme demonstrates the mutual relationship between causality ("cause leading to consequence") and finality ("required consequence leading to identified cause"). If the contents of figure 5.13 are interpreted in reverse direction (from right to left) we obtain the causality: "If the TS is of this type, then it exerts this definite effect". This process of "if-then" assertion is repeated until the last effect in the action chain is reached, which is for engineering purposes the final effect (*causa finalis*).

The aims-to-means procedure described above can be refined by inserting various category-forming characteristics between the aims and the means. That these are defined in the theory of technical systems as *design characteristics* should not be surprising once we have recognized their key position.

This description of the search for a suitable TS (the means) to fulfill the needs (aims) can be termed the "basic algorithm" [108] of the design process.

Concerning the TS structures, the following "why - with what?" steps give a connection to the terminology used in these discussions:

Goals:	–	Means:
change of the operand	–	technologies (partial transformations)
technology	–	effects
effect (purpose function)	–	inputs, (technical) functions
(technical) function	–	organs
organ	–	constructional elements

Between these goals and means, questions must be asked at each level about states of embodiment of the design characteristics that lead directly or iteratively to the means.

5.7 TS Boundary

The boundary of a technical system delimits its functioning unity, and must also be regarded as delimiting its spatial unity. At the surface of the TS there are boundary elements that mediate the connections to the environment, the inputs and outputs. These boundary elements are termed *receptors* and *effectors*, which contain the action sites of the technical system (where the TS exerts its effects or receives actions from outside). In all diagrams of TS structures, the effectors and receptors are represented directly at the system boundary (compare figure 5.4).

5.8 TS Environment

As may be seen from figure 5.4, the following interdependent and interacting systems are always present in the vicinity of a technical system (compare figure 3.1 B):

- technical process with operand;
  (Under certain circumstances only the "operand" is present as receiver of the effects of a TS in its environment. The term "technical process" refers to the application of technical systems and describes it from various aspects.)
- human (individual or group) in direct contact with the TS and participating in or regulating the effects exerted onto the TP;
- technical systems that in some way assist the technical process in transforming the operand, usually by exerting their effects;
- active environment, that part of the total environment that has a direct relationship to the TS being considered. The active environment also contains the "fixed system" (mainly our planet Earth) with which the TS must be connected. The laws of nature can also be counted as part of the active environment, therefore the TS exists within the regions where these laws are active, which corresponds to reality.

Whether an individual component or element should be regarded as part of the TS environment, can and must be resolved with the help of the appropriate transformation system, particularly the transformation process. Such decisions sometimes present severe problems. The TS environment is therefore dependent on how the transformation system is defined. It is evident that a TS can belong to several transformation systems, and may therefore need to perform in different environments. In addition, its environment will change during the various phases of origination - design, manufacture, testing, commissioning, etc. (compare figure 5.2 and chapter 10). The TS must be compatible with (and capable of withstanding) all these types of environment.

5.9 Technical Systems - Statements

The insights about technical systems can be summarized as follows:

Statement 5.1 The task (aim, purpose) of technical systems consists of exerting particular effects on the operands in the technical process. Consequently, the TS must have the ability to exert these particular effects.

Statement 5.2 The effects are exerted from a system of action sites - effectors - (at the boundary of a technical system) under certain action conditions (compare figure 5.7).

Statement 5.3 The effect is achieved as the output of an action process (through an action chain) internal to the TS, in which the input measures are converted into the effects (output measures) of that TS. The action process is a direct consequence of the structure of the TS, the term "mode of action" of the technical system is used to describe this relationship. The action process takes place after it has been initiated at the desired moment.

Statement 5.4 The action process (and the mode of action) is characterized by the TS-internal conversion of inputs of material, energy and information. These internal conversions are termed (technical) functions, and are linked by their relationships in the function structure of the TS.

Statement 5.5 The (technical) function describes the capability of the TS to transform an input quantity into an output quantity of different measure and/or form by natural phenomena that are deliberately caused to take place.

Statement 5.6 The input quantities to a technical system are in the forms of materials, energy and/or information.

Statement 5.7 Technical systems exist during their origination, operation and disposal phases in various roles and states, according to the transformation systems within which they belong at that time.

Statement 5.8 In general, technical systems possess specific relationships to all other elements of the appropriate transformation system. Apart from the desired inputs and outputs, various other (undesired) inputs and outputs exist that are termed disturbances.

Statement 5.9 The functions (and function structures) are realized by organs (and organ structures) using certain action principles. Organs are the means employed to fulfill the aims of the functions. Each function (function structure) can be realized by several existing action principles.

Statement 5.10 The organs (and organ structures) are realized by constructional elements (and component structures). The component structure realizes also all types of properties, as will be discussed in chapter 7. Each organ (and organ structure) can be realized by several available component structures.

Statement 5.11 Each of the TS structures depends on the state of embodiment of certain design characteristics. This is the cause for the alternative TS structures that are possible.

Statement 5.12 Each function, each organ or constructional element may be resolved into a system of partial functions, partial organs or partial constructional elements, resulting from the hierarchical character of technical systems.

6 Classification of Technical Systems (Systematics)

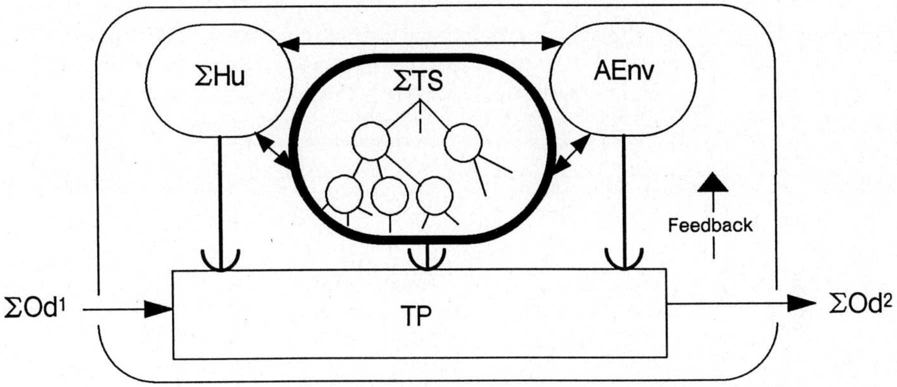

The abstract term Technical System (TS) was selected to represent all technical products (see chapter 1). Technical systems realize (by exerting their effects) the transformation of operands in many processes. The range of application for technical systems is very broad, and includes all branches of the economic system. Figure 6.1 shows a wide variety of technical systems that are available. The reader is encouraged to find further examples in banking, the stock market, government, defense, the home, recreation, sports, and any other human activity.

Classifying these technical systems according to similarity and relationships helps to bring order into the vast family of technical systems, and makes an overview possible. Classification also makes it possible to transfer progress and experiences within and between the categories (labelled "technology transfer"), often providing new insights into both categories. The process of classifying can also produce new insights, and the classifying criteria may fulfill a special mission. These criteria should first be listed, and characteristic examples shown.

Technical systems can be classified in various ways, resulting in a multi-dimensional matrix. Some of the more useful classification criteria are as follows:

- according to the function (and purpose) of the technical object, e.g. TS for holding, deforming, turning, lifting, ... ;
- according to the type of operand, e.g. TS for transforming materials, energy, information, or biological objects (including humans) as operand;
- according to the type of process they drive, e.g. processing, manufacturing, transporting, storing (compare figure 4.11);
- according to the principle by which the working effects are achieved, e.g. TS with mechanical, hydraulic, pneumatic, electronic, chemical, optical, acoustic, ... principles, or a combination of these;
- according to their conditions of behavior, e.g. TS classes (graduated steps of numerical values) of power or speed, TS for various dimensions of operand, TS for various environmental conditions (tropical or arctic climate, desert, etc.);

Branch of Economy	Technical System TS	
	Equipment for	Typical Machine
Mining	Accessing Delivering Preparing	Cutting machine Conveyor Screening machine
Energy generation	Steam raising Electric generating	Steam boiler Water conditioner Steam turbine Gas turbine Water turbine Generator
Smelting	Pig iron smelting Steel smelting	Blast furnace Bessemer converter LD oxygen processor Rolling mill
Chemical industry	Coal scrubbing Color producing Explosives producing	Pressure vessel Piping Destillation column
Metalworking industry	Chipless forming Chip-forming Heat treatment Foundry Assembly	Press Forging hammer Machine tool Furnace Forming machine Jigs and fixtures
Constructional industry	Oil exploration Building Roadworks Hydro-construction Materials manufacture	Drill rig Personell lift Scraper Concrete mixer Block press
Transportation	Railway Shipping Space travel	Locomotive Wagon Passenger liner Rocket
Textile industry	Textile manufacture Dressmaking	Spinning machine Weaving loom Sewing machine
Food industry	Sugar refining Cheese production Milk processing	Concentrator Press Centrifuge
Medicine	Diagnosis Therapy	X-ray apparatus Artificial heart Prosthesis
Printing, offices	Printing Office work	Printing machine Typewriter Calculator
Agriculture	Transporting Harvesting Lumbering	Tractor Combine Chain saw
Distribution, trade	Self-service Packing	Check-out Wrapping machine

Fig. 6.1 Examples of Technical Systems for Important Branches of the Economy

- according to the degree of complexity, e.g. TS as component (part), machine element, sub-assembly, assembly group, machine, plant, ... ;
- according to similarity of manufacturing processes, e.g. classes of TS made by casting, forging, pressing, turning, ... ;
- according to the difficulty of designing the TS;
- according to their form, e.g. as rotational elements, plane elements;
- according to the arrangement within the TS of systems of lower levels of complexity, e.g. manufacturing plant arranged according to machine type (functional grouping of machine tools) or with closed technological cycle (flexible manufacturing cells, group technology, or dedicated flow-line);
- according to material of manufacture, e.g. made of steel, brass, plastics, reinforced concrete, silicon, germanium, ... ;
- according to the degree of originality of the designed object, e.g. re-used (drawings, etc. from a previous different application), adapted, re-designed, newly designed (and developed) TS;
- according to the type of manufacture, e.g. TS in single items (one-off) manufacture, batchwise, series, mass-production;
- according to the manufacturer, e.g. TS made by Siemens, Ford, Boeing, IBM ... ;
- according to other properties of the TS, e.g. operational, appearance, delivery, economic, ... ;

It is obvious that each technical system is simultaneously a member of a large number of categories, and in some cases can not be uniquely fitted into any single category within a classification.

Some of these categories of technical systems are discussed in more detail in the following sections. In particular, those categories and criteria that are important for the engineering designer or project engineer are discussed.

6.1 Classification of Technical Systems by Function (Effect)

Technical systems are often given names according to their function. The products of mechanical engineering are usually classified and listed using one or more of their TS functions as criteria. These are generally connected with application, sales, planning, and general overview of available items. It would be very useful if the classifications, and the terminology used within them, were common to all fields of engineering.

Designating a technical system according to its function is especially appropriate where buyers are searching for a means to fulfill a certain purpose and function. This should make orientation and the buying task easier, and applies to terminology and categories that should be used in company catalogs, survey tables, etc. Examples of surveys may be found in exhibition catalogs, and product comparisons by trade journals.

Machine elements and sub-assemblies are also regarded as technical systems according to the theory of technical systems. Since the designer must use these elements in applications that suit their capability they should also be classified according to their function. This type of classification is termed "design-functional". It provides easily digested design information about existing TS, and enables re-use of TS (as already designed or manufactured parts that are available, standardized or standard) in other applications, unification, and standardization of components and constructional groups. The designer can save time if suitably categorized lists are available.

A unified classification of parts, elements and constructional groups (e.g. subassemblies) can only be achieved with difficulty in most fields of engineering, especially as a central facility or "data bank", because the range and variety of items is generally too large.

Each company should produce its own classification scheme and index, if possible following the above general principles. Every enterprise uses a number of parts and constructional groups which fulfill a common function in their field of engineering, e.g. fasteners, gear boxes, couplings, pneumatic cylinders, hydraulic motors, and so on. Many other parts or groups have specialized functions, such as measuring equipment, regulating and signalling devices, hydraulic and pneumatic valve and control elements and their components, special electrical and electronic accessories, jigs and fixtures, and many more. To include all these in a unified classification system would make it too cumbersome and confusing.

When setting up such a classification of components and groups it is important that the labelling codes (number and letter combinations) should be simple, otherwise memorizing them would be too difficult. A three or four character code seems to be best, unless the label is easy to interpret as a mnemonic. This usage can be compared with the common systems of stating telephone numbers, generally as seven-digit numbers quoted in North America in groups of (3+4), and in parts of central Europe in groups of (2+2+3). Studies in psychology suggest that our short-term memory can only handle seven items or concepts (plus or minus two) at any one time, but grouping can help memory.

Titles of parts and constructional groups on detail or assembly drawings should preferably be chosen to indicate their functions, rather than their form. This eases classifying new parts into an index according to design-functional guidelines. For instance, an annular seal should be given the name "seal" rather than "ring". The term "sealing ring" may be too complicated for this purpose, and not unique enough for a user to be able easily to classify a "ring seal" into the same category, unless a listing of synonyms is available.

6.2 Classification of Technical Systems by Action Principles

It is important for the designer to classify all technical systems of similar function into groups according to the additional common characteristic, the *action principle.* For example, the technical systems "industrial furnace" can be classified according to their source of energy: electrical, gas, solid-fuel, and liquid-fuel. Electrical furnaces can in turn be sub-classified according to the (physical or action) principle used to convert the electrical energy into heat, such as resistance (direct or indirect heating), discharge (electric arc) and induction. According to the mode of action we can distinguish between furnaces with continuous feed and those with intermittent work cycles (chamber furnaces). The continuous feed furnaces can again be divided according to the action principle used to transport the goods to be heated (the operand), such as roller, conveyor belt, conveyor chain, pusher, and stepper-linkage.

These classifications are typical for technical systems, the group of characteristics used as classifying criteria are termed the *functionally determined properties.* They are very important to the designer who uses systematic approaches and methods.

6.3 Classification of Technical Systems by Degree of Complexity

Categorizing technical systems according to their structure (compare figure 5.4) is already usual and essential for the purpose of engineering design. The principal characteristic for formulating such categories is their function, as explained in sections 6.1 and 6.2. This classification permits re-use of a drawing or group of drawings for an application that is different to the one initiating the design work. The needs of production dictate a different set of classifying criteria, for instance from the viewpoint of assembly. It may be advisable for organizational reasons to divide a technical system into smaller, less complex partial systems, which are only able to fulfill their functions after assembly with other subsystems.

A survey of classifications of technical systems according to their degree of complexity is given in figure 6.2. The relationships between these degrees of complexity is illustrated in figure 6.3. Further intermediate scale points could be iden-

Level of Complexity	Technical System	Characteristic	Examples
I (simplest)	Part, Component	Elementary system produced without assembly operations	Bolt, bearing sleeve, spring, washer
II	Group, Mechanism, Sub-assembly	Simple system that can fulfill some higher functions	Gear box, hydraulic drive, spindle head, brake unit, shaft coupling
III	Machine, Apparatus, Device	System that consists of sub-assemblies and parts that perform a closed function	Lathe, motor vehicle, electric motor
IV	Plant, Equipment, Complex machine unit	Complicated system that fulfills a number of functions and that consists of machines, groups and parts that constitute a functional and spatial unity	Hardening plant, machining transfer line, factory equipment

Fig. 6.2 Technical Systems Classified by Degree of Complexity

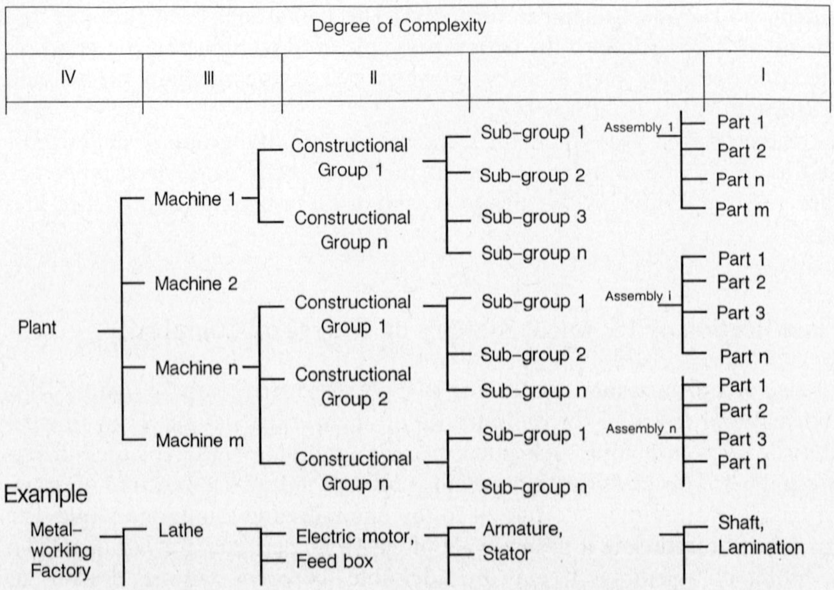

Fig. 6.3 Relationships between Levels of Complexity of Technical Systems

tified, especially in the technical systems of higher complexity. Any TS group can be further divided into subgroupings. We should recognize that the stated degrees of complexity are relative within each hierarchy of TS. For instance, a "plant" can be regarded either as a totality of TS for an industrial operation, or a grouping of technical systems (e.g. machines) for an application within the total plant, or an operational unit (a single complex machine), or a basic unit (an operational module). In one transformation system, a particular technical system at a lower level of complexity can be regarded as a subgroup or even as a purchased, bought out or out-sourced component (e.g. an electric motor), whereas in another transformation system it is the state of finality (e.g. for the manufacturer of motors).

It is generally accepted that technical systems of lower degree of complexity can find a greater breadth of application and more versatile use. A threaded fastener, for instance, can be found in most engineering products, an electric motor in many, but a transfer-line of modular machine tools can only be used for one special process.

Classification under this criterion is important to the design engineer, because the degree of complexity of technical systems:

- is directly correlated with the degree of complexity and difficulty of finding a solution to a design task;
- indicates certain limits of specialization for engineering design activity, e.g. a project engineer will deal with designing a whole plant, a design engineer with a machine for that plant, and a detail designer with the constructional elements;
- provides orientation for the designers' work, because when working at a defined level of complexity they need only acknowledge that the results of this task

must be incorporated into the higher levels, and for transfer to the lower levels only the conceptual problems need to be solved.

Each level of complexity of technical systems also represents the appropriate manufacturing and assembly totalities, based on their assembly drawings. Particularly for constructional groups, sub-groups or sub-assemblies, formulating appropriate totalities is necessary as prerequisite to modular construction of technical systems, and provides a basis for systematic organization of the production processes (e.g. defining an assembly sequence).

6.4 Classification of Technical Systems from Manufacturing Similarity

Certain groups of technical systems need similar manufacturing equipment for their production. Examples are producing steam boilers and chemical process equipment in the same fabrication shop, or manufacturing lathes, milling machines, boring machines and other machine tools in the same workshop facilities.

In the same way it is possible to classify machine parts according to similar technologies of processing, whereby the external form (mainly shape, dimensional tolerances, geometric tolerances and deviations from ideal geometry, and surface finish) is the guiding characteristic.

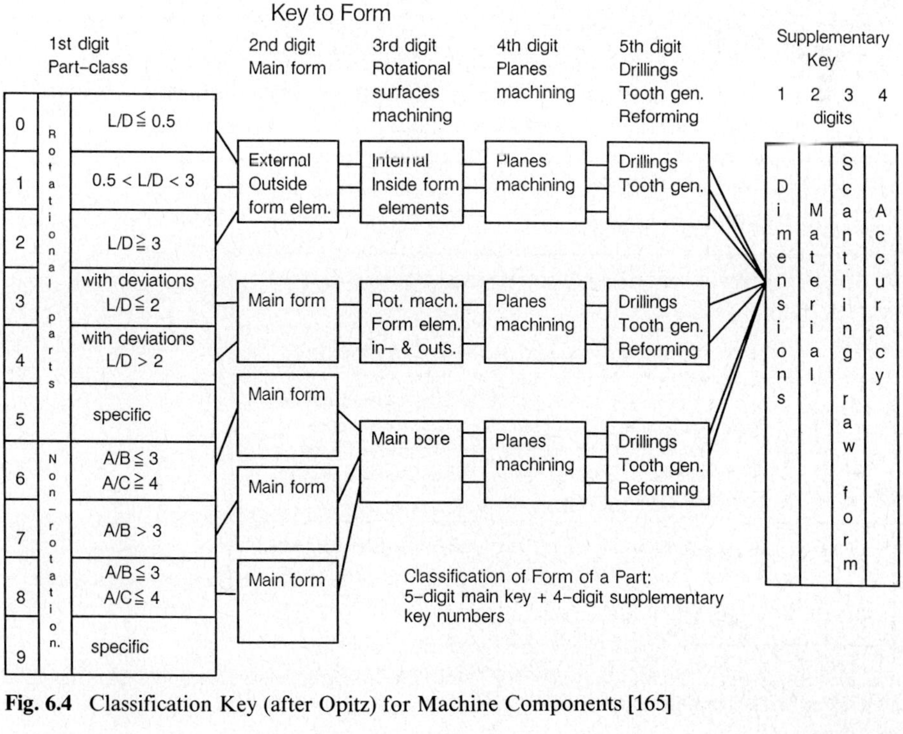

Fig. 6.4 Classification Key (after Opitz) for Machine Components [165]

Such classification enables rational and technological preparation for production. It can also lead to better organization of the production processes, for instance by suitable space arrangement of the workplaces or machines for manufacturing similar parts. This can lead to various rationalizing actions, for instance transferring a production program into a manufacturing facility with a related and specialized technology.

This classification criterion is important for people involved in work preparation, factory shop-floor management and planning. It is a part of the concept of "group technology" and "flexible manufacturing". These concepts are implemented by placing suitable machines into work cells which can complete the manufacture of a family of constructional elements in one physical area of the facility. Transportation between different types of machine tools is avoided, and consequent storage times and inventory of intermediate products are reduced. One possible classification scheme for mechanical components is outlined in figure 6.4. Parts with similar codes are sufficiently similar to be made in one work cell.

6.5 Classification of Technical Systems by Difficulty of Designing

Technical systems can be categorized into different groups according to the relative difficulty of designing them. As an example, figure 6.5 shows technical systems of the third degree of complexity (refer to figure 6.2) divided into six groups of design difficulty. Depending on this degree of design difficulty an expert with suitable knowledge, skills and abilities must be chosen to perform the design work. Design difficulty is also a criterion for planning purposes, and may be used to establish an estimated time to complete the design work.

Degree of Design Difficulty	Examples
I	Small containers, simple supporting structures, troughs, control panels
II	Complicated and larger containers, valves, simple couplings
III	Pressure lubricators, trailers, heated process baths, destillation apparatus, throttle valves, floor–controlled bridge cranes
IV	Special ventilators, heat treatment furnaces, pressure vessels, centrifugal pumps, personnel lifts
V	Gas generators, tower cranes, high voltage transformers, piston pumps
VI	Metalworking automats, planetary gear boxes

Fig. 6.5 Examples of Technical Systems of Third Level of Complexity (see figure 6.2) Classified by Degree of Design Difficulty

Degree of Design Difficulty	Characteristics	Examples
I	Very simple parts with few co-ordinates (dimension arrows)	Washer, simple lever, small shaft, bolt, clamp
II	Simple parts with many co-ordinates	Lever, simple gear wheel, bearing
III	Complicated parts	Gear wheel, spline shaft, complicated pressing
IV	Complicated parts	Complicated casting, small forging
V	Very complicated parts	Complicated casting of box form, medium-sized forging
VI	Very complicated and large parts	Machine base, machine housing, welded or cast machine mount
VII	Most complicated parts of large dimension and unusual form	Turbine blade, large forging

Fig. 6.6 Machine Components (First Level of Complexity according to figure 6.2) Classified by Degree of Design Difficulty

Machine parts (first degree of complexity according to figure 6.2) can be divided into groups according to design difficulty. An example of such a classification into seven groups is shown in figure 6.6.

Criteria for assessing design difficulty are: required originality; complexity of the functions, the forms and the structures of the TS; difficulties in modelling and analysis (calculation); physical size of parts; number of dimensions; particularly onerous requirements with respect to properties and their optimization, such as minimum weight, manufacturability, low costs, appearance, and so on.

6.6 Classification of Machine Parts and Groups (Sub-Assemblies) by Production Location and Degree of Standardization

This classification is very important for assessing the economics of a technical system. Information about the possibilities of small-batch, large-batch, flexible or continuous manufacture is provided by the proportions of repeated groups within technical systems.

A survey of the structure of a higher technical system with respect to its groups and parts according to this classifying criterion is shown in figure 6.7.

The following relationships are valid:

(a) total number in categories according to figure 6.7:

$$n = n_D + n_{Re} + n_{Typ} + n_{WSt} + n_{St} + n_{PuD} + n_{PuTyp} + n_{PuSt}$$

and (b) relative numbers in categories:

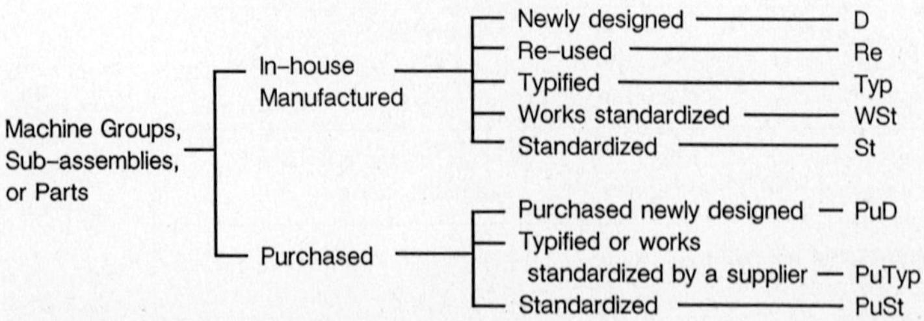

Fig. 6.7 Structure of Groupings and Parts from Manufacturing Location and Degree of Standardization

$$1=\frac{n_{D}}{n}+\frac{n_{Re}}{n}+\frac{n_{Typ}}{n}+\frac{n_{WSt}}{n}+\frac{n_{St}}{n}+\frac{n_{PuD}}{n}+\frac{n_{PuTyp}}{n}+\frac{n_{PuSt}}{n}$$

where n ... the total number of all parts or machine groups,
n with suffix ... the number in each category.

A number of different procedures can result in standardizing:

- international agreement, e.g. the SI standards of measurement, ISO screw threads, ruling sizes and tolerances for ball and roller bearings;
- national agreement, e.g. codes of practice for concrete, connection dimensions of electric motors;
- agreement by industrial product, e.g. AGMA standards for gearing, operating systems for personal computers;
- company in-house agreement, e.g. selection of material sizes to be stocked;
- market trends, e.g. communication software and hardware.

The first two categories are binding agreements that may or may not be enforceable by law. In most countries a national standards institution is responsible for obtaining agreement on the contents of each standard, a process that involves negotiation between representatives of industry, education, government, etc. Some countries have more than one institution working on standards, a national institute then coordinates their efforts.

Any TS components that result from the first two of these procedures are usually termed "standard parts". Results of the other procedures are frequently labelled "standardized". The main purpose of standardization is to control quality and interchangeability of TS or some sub-groups.

In the case of company practices, one of the dominant trends is to rationalize families of similar systems into a smaller number, permitting larger manufacturing quantities for each. This process is also referred to as variety reduction, or typifying.

The market trend is established by the marketing power of the primary supplier, and willingness of secondary suppliers to follow. The examples of preferred operating systems for micro-computers, such as CP/M for eight-bit machines, MS-DOS for 16-bit computers, and UNIX, illustrate how a de-facto industry standard may emerge, even without formal agreement.

From an economic point of view, the proportions of newly designed inhouse (category "D") and bought out (category "PuD") items that exist in a technical system should be as small as possible. They reflect the requirements for designing and preparing for manufacture. A minimum proportion of items in these two categories "D" and "PuD" are also prerequisites for quantity manufacture because the existing parts can thus be manufactured in larger quantities. Other considerations may justify why this viewpoint can not be used.

In any case, every report about a completed design task should contain information on the proportions of parts, components and sub-groups in each of the classifications shown in figure 6.7.

6.7 Classification of Technical Systems by Design Originality (Degree of Novelty)

A combination of the economic considerations mentioned in section 6.6, reducing risks (not being able to design to fulfill the functions, or failure of the TS, or hazards), increasing reliability, and other reasons force the designer to retain as many as possible of the previously designed and tested items and completed technical systems. Technical systems can be classified according to their originality or degree of novelty into the following categories:

- *re-used* technical systems. A set of technical systems, or a set of machines or parts exists or is available to fulfill the required functions, and the most appropriate ones can be selected and used without alterations. This group comprises standard parts such as screws, keys, valves, springs, etc., and standardized parts such as purchased (bought out or out-sourced) components and assemblies. It also contains non-standardized parts and groups that can be re-used in a different context from the application for which they were originally designed.
- *adapted* technical systems. This is a type of variant design. Some technical systems may exist for the required functions, but they do not comply with all requirements of that application. Some alterations must be made in terms of size, power, speed (rotational or other), connective or other dimensions, or material or manufacturing methods. New materials are generally only used to improve the quality of the technical system, to reduce costs, to modernize the system, or to adapt it to other operating or environmental conditions (e.g. corrosion reduction). The main concepts, structures and other important properties of the TS remain unaltered. The technical system is merely adapted to the particular conditions and requirements of the new task. Even this adapting may prove under certain circumstances to be difficult.
- *re-designed* technical systems. This is generally a more difficult type of variant design involving some conceptual changes. The existing technical systems do not completely fulfill the requirements with respect to their functions and other properties. In the re-designed version, only the functions, some of the parameters, and possibly the working principles remain unaltered. Parts may have their form, dimensions, material or technology revised. In more complex technical systems the alterations involve their construction and their structures, i.e. parts and constructional groups, their couplings and spatial arrangements. Re-designing is probably the most frequent design activity.

- *original* technical systems. No existing technical system can fulfill the required functions, or the existing system has distinct deficiencies. A new working principle and additional technical properties are usually needed. As an example, a design task may be stated as: the effect "heating" which was accomplished by a technical system using the action principle "burning a mineral oil derivative" should be replaced by a newly designed system using "electrical resistance heating".

6.8 Classification of Technical Systems by Type of Production

The type of production depends on the number of nominally identical items to be produced, and gives each product some characteristic technical and economic properties.

- *individually manufactured* (one-off) technical systems. In these TS, the work and cost of designing and preparing for manufacture must be absorbed by a single technical system, with a resultant increase in its cost (usually also reflected in its price). The risk of whether the required functions can be achieved at all is generally high, because for larger technical systems the manufactured item (or production unit) is simultaneously the prototype and the technical system in its state of finality, which is therefore intended for continuous use. This category makes the greatest demands on the engineering designer. The task is especially difficult if the action principles are also relatively untried, as was the case for instance for the early nuclear reactors, and the first supersonic aircraft.
- *series or quantity manufactured* technical systems. These include large batch, and mass produced systems (continuous or flow-line production). These technical systems are generally more carefully worked out in the design stages, and more completely prepared for manufacture. The larger number of identical items to be produced results in a lower proportion of design costs per item. The improved design and manufacturing preparation is reflected in all properties of the technical system, including its appearance. Verification, checking and testing is usually performed on a relatively limited sample of the produced items (and even then human error can overlook some obvious faults), therefore manufacturing faults are not excluded in such technical systems. Only if a complete and faultless verification and inspection of all operations, parts, sub-assemblies, and the finished product could be carried out would it be possible (but expensive) to achieve uniformly high quality. Some computer-managed systems are able to approach this ideal.

The expert can use these categories of production types as a basis for assessing the probable quality of a technical system.

There is a clear tendency toward increasing the standardization and series-production of technical systems, particularly for various functions of low degree of complexity, as for instance in the fields of connectors, fasteners, measurements, regulation and controls, prime movers (drives), and distribution. Simultaneously, the number of specialized one-off manufactured technical systems are also increasing, because modern production depends on special equipment, single-purpose tools, automata, and transfer or flow lines. These one-off systems permit cheaper manufacture of standard products. Both categories place ever greater demands on the volume and quality of design work.

An intermediate category has recently surfaced. It permits greater variety in the products by using specialized equipment in the form of computer-aided manufacture and flexible manufacturing cells. Preprogramming of versatile computer controlled machine tools (including CNC or DNC "machining centers"), coupled with computer-controlled feeding and positioning of workpieces ("robotics"), allows the production of a variety of small batches with only minimal disruption from change-over between batches. The main source of this versatility lies in providing software, particularly the manufacturing programs, and by applying concepts such as JIT (just-in-time) delivery of components for processing, CIM (computer-integrated manufacture), etc. Similar developments have occurred in assembly, usually using robotics, which allow a "mix-and-match" variation of the assembled product from item to item, with pre-programmed delivery of the necessary parts. The automotive industry is probably a leader in this field.

6.9 Classification of Technical Systems by Degree of Abstraction

Technical systems are one of the most numerous classes of objects in the world, and can be sub-divided in a hierarchical manner. Figure 1.1 shows the start of such a categorization. Other classes can be formulated using a systematic taxonomy similar to that of biology. These are based on various aspects of similarity and morphological characteristics, namely kingdom, phylum, class, family, genus, species, etc.

An examination of the various forms of TS structure (sections 5.3 to 5.6) can be used as a basis for this categorization. Focussing on the *kingdom* of technical systems yields the hierarchy of categories shown in figure 6.8. These categories and their hierarchy are illustrated by an example from the phylum of "machine tools". The possibility of allocating the various TS structures (compare figure 5.4) to the hierarchical steps shown in figure 6.8 may be surprising, but clearly demonstrates the central role of the design characteristics in the definition of the various levels of abstraction.

6.10 Classification of Technical Systems by Type of Operand

The known types of operand - biological and non-biological material, energy, and information (commands and data) - can be used to make a further classification in the same way as used for the technical processes (compare figure 4.11). Thus we can distinguish technical systems that convert:

- material,
- energy, and
- information.

Earlier this viewpoint led to dividing technological equipment into various specific areas of activity, such as production (material-processing) machinery, motors and generators (energy-processing), communications (information-processing), etc.

Level of Concretization	General Hierarchy			Example	
	Designation of Level	Defined by (Graphical Model)	Established Design Characteristics	Designation of Level	Graphical Representation
0	Machine System (MS)	TS Σ Ef (input) Od^1 → Trans → Od^2	MS with mainly mechanical mode of action		
0.2	Phylum of MS	General Black Box Family of MS Od^1 → Class of Trans → Od^2	– Phylum of Operands – Class of Transformations	Machine Tool	Machine Tool Rigid Body Form 1 → Forming Process → Rigid Body Form 2
0.4	Class of MS	Detailed Black Box Sketch of the Technological Principle Basic Function Structure	– Family of Operand – Technological Principle of Transformation – Necessary Input Effects and thereby the basic Functions	Metal-cutting Lathe	Lathe Rigid Body Rotational Form 1 → Forming Process → Rigid Body Rotational Form 2 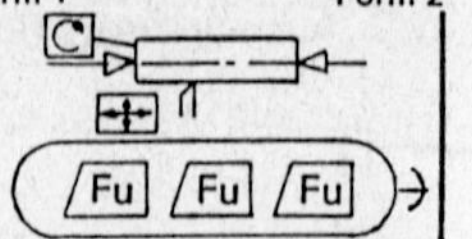
0.6	Family of MS	Detailed Function Structure Rough Component Structure Concept Sketch	– Species of Operand – Function Structure (Functions and their Sequence) – Inputs to MS – Families of Function-carriers (Organs) – Combination and Basic Arrangement of Function-carriers	Universal Metal-cutting Lathe	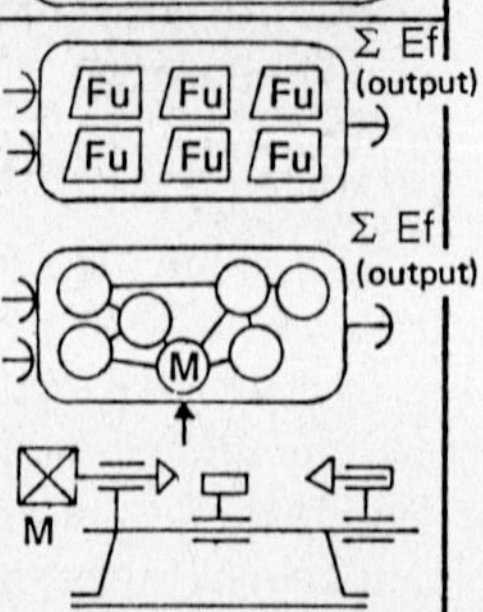
0.8	Genus of MS	Drawing of Component Structure (Similarity conditions differ for different Type Series) Drawings of Common Components and Subassemblies	– Complete Parts – Arrangement – Partial Form – Some Dimensions – Types of Materials – Some Tolerances and Surface Property Definitions	Lathe type SUR	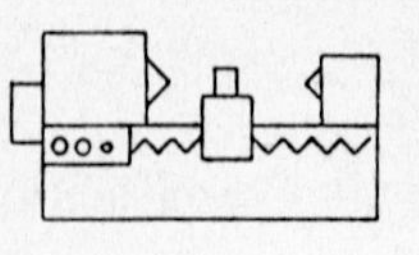
1	Species or Serial Size	Complete Set of Workshop Documentation	Total and Definitive Specifications for Parts and Arrangement. For all Parts: Forms, Dimensions, Materials, Manufacturing Methods, Tolerances, Surface Properties	Lathe Serial SUR 5	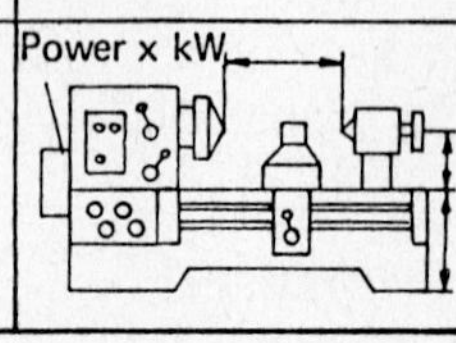

Fig. 6.8 Levels of Abstraction of Technical Systems and their Representation, Example of Phylum Machine Tools (Reproduced from V. Hubka, Principles of Engineering Design, London: Butterworths, 1982 [119] by kind permission of the publishers)

The theory of technical systems presented in this book regards the TS model as a system to deliver desired effects. The categorization from type of operand is thus secondary, and may even be misleading by implying that the application of a technical system is restricted to that range of operands.

6.11 Classification of Technical Systems by Application in the Technical Process

Discussing the transformation systems and its elements in chapters 3 to 5, it was mentioned that not all technical systems can be applied directly as deliverers of effects to the operand in the technical process. This classification is generally not connected to criteria described in other sections of this chapter, especially to complexity. The classifications consist of:

- final TS that deliver effects to the operand in the process, and
- internal TS that deliver effects to other TS.

The final systems are the ones that are recognizable as the direct operators of the technical process. These are usually termed plant, machine, apparatus, and also tool and instrument. They form an active pair with the operand, and exert the output effects of the TS.

In contrast, the internal technical systems such as gear boxes, linkage mechanisms, electrical and machine elements, and so on, exert their effects on other TS. They are usually labelled "sub-systems". Such TS apparently have other TS as their "operands", which is generally not permitted in our conventions. They can only have cooperative associations with other such technical systems, and do not exert any effects onto the operand of the transformation process. They only form a working unit as a final TS in aggregate with other internal TS.

6.12 Classification of Technical Systems by "Quality"

Considerations of quality (a measure of total value or partial value) will be treated in chapter 8. The classes of quality for consumer products are generally well defined and enforced. For most technical systems the classes of quality are generally selected *ad hoc* for evaluations to fulfill certain purposes.

6.13 TS Systematics - Statements

The insights developed in this chapter are summarized as follows:

Statement 6.1 Technical systems may be classified according to various characteristics that they have in common, seen from different points of view. The set of all technical systems can be divided into a number of different sub-sets according to various criteria. The resulting categories can serve various purposes, such as creating order, overview, cataloging, evaluating, characterizing, understanding, comparing, etc.

7 The Properties of Technical Systems

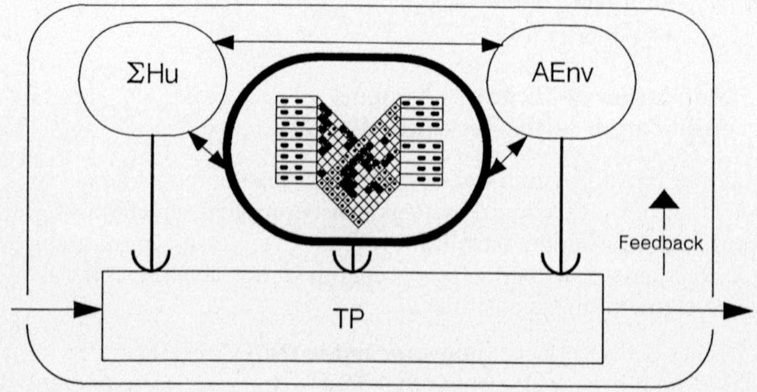

Humans investigate and explore their environment by asking many questions, as the main way in which they to try to find patterns and structures. In the words of a famous poet:

> I keep six honest serving men
>   (They taught me all I knew);
> Their names are What and Why and When
>   And How and Where and Who.
> (R. Kipling, "Just So Stories", from The Elephant's Child)

One of the most frequent types of question that the human being asks starts with or implies the word "how". Typical questions of this kind are: "What is the nature of the car?" – or "How is it constituted?" or on a more personal note "How do you like it?"; "How fast does it go?" – at present, or at maximum; "How much fuel does it use?" – or "What is its fuel consumption?"; "How does it look?" – or "How good is its appearance?"; "How is its color?" – or "What color is it?"; "How does the car work?"

The answers may be: The car has a maximum speed of 120 km/hr and a fuel consumption of 0.12 ml/m (or more usually 12 l/100 km). Such statements give unambiguous information to everyone familiar with the measurement system. Alternatively, the answers may be: The car moves quickly, it has low fuel consumption. This characterization is more ambiguous because it is relative. It is not clear whether "quick" is intended to mean 20 km/hr or 200 km/hr. To avoid this we can ask the questions in more precise terms: "How many kilometers per hour does this car travel?" Questions that ask or imply "how many" or "how much" are very important, because they ask that a specific *property* of a technical system be *quantified.* And, assuming that someone knows the answer, the *measure* (numerical value from an appropriate scale) of the property is precisely determined (within the limits of accuracy of the measurement system used).

A more difficult form of question to answer is: "How reliable is this vehicle?" The answer can be qualitative: "it is reliable" or "it is not reliable". Such an answer delivers a relative characterization that depends on the viewpoint of the assessing person. A person whose car fails only three times during six months will declare it more reliable that the car that had to be repaired every week. A definition of reliability can be formulated, and quantification could be achieved by suitable experiments, giving an answer to "how much". This is the role of the concepts of reliability and maintainability, which have been defined on a probabilistic (stochastic) basis, giving information about average (and extreme) failure rates over suitable time periods.

A still more difficult question is: "How does the car look?" We hear the answers: It is beautiful, ugly, elegant, etc. This is a very unclear evaluation, and it is not easy to quantify such a variable in time and space – it is subjective and qualitative. It would be adequate if the personal opinion of one ideal person could be used as the basis of a measurement, this one subjective opinion would form an objective scale for everyone else. This is done in a way by averaging a number of subjective assessments (e.g. in opinion polls), using an arbitrary scale.

Selecting the properties by which to characterize a technical system is another problem. A large number of questions can be asked about any object. For instance, other questions about the car could be: What is its compression ratio? At what rotational speed does the motor run? Where is the final drive, to the front or to the rear wheels? These types of question are likely to arise in conversations between specialists, because the quantities (measures, units and properties) are defined in terms that they understand, and they know what they can expect if told that a car's motor has a compression ratio of 10:1, runs at 7000 rev/min, and the car uses front-wheel drive.

These examples show the scope of problems generally connected with the properties of technical systems (in this chapter) and their evaluation (in chapter 8).

Every technical system has a purpose (or a system of purposes consisting of individual purposes and their relationships), it is intended to exert some desired effects on the operands of a transformation process. Every technical system exhibits a behavior which is determined by its mode of action and its structure (a system of elements and their relationships). The intended behavior is called the system's function. Every technical system accepts inputs (a) from humans, (b) from other technical systems and (c) from the active environment. Some of these inputs are desired, to initiate or control the intended effects, others are secondary inputs acting mainly as disturbances. Each technical system has outputs, which include the desired effects, and the undesired secondary outputs. These desired and secondary outputs influence (a) the operands of the process, (b) humans, (c) other technical systems, and (d) the environment. Every technical system passes through a succession of distinct states, in the course of normal operation, and during its life.

Every technical system, and each element and relationship at or within the boundary of a technical system, has certain *properties* that belong to the system and define it. They include size, weight, form, speed, stability, suitability for manufacture, reliability, and so on. There are no objects without properties. It is not only the property that matters, but also the form in which that property appears –

its *state of embodiment*, and its *measure*, value, quantity and/or quality. These properties and measures distinguish similar systems from one another.

Many of the properties are more or less objectively measurable, others can only be assessed by subjective means. The state of the technical system at any one time is defined by the total of all its properties and their state of embodiment (their measures). The principal property of any technical system is its *capability of doing something* (preferably useful – its function), and is reflected in its behavior.

A *property* is thus defined as any characteristic of an object, in our case of the technical system, that belongs to and characterizes it. A property in this sense is for instance the capability to behave in a specific manner, or to exert a specific effect. These properties can be used to characterize a technical system. An objective characterization and assessment of a TS requires that the properties used as evaluation criteria are quantifiable.

This broad interpretation of the term "property" may cause some difficulty for the engineer, because it includes function, capability, dimensions, maintainability, transportability, manufacturability and many others. Examining the implications will show that this broad interpretation is correct (consider the further explanations about systems and properties in appendix B).

We must remember that the most important aspect of a technical system is its set of desired properties. The technical system is only the *carrier* of the necessary functions and the desired behavior, it is not usually a purpose in itself.

A further set of properties is linked to the function of a TS: suitability for operating (internal actions) and being operated (by inputs from the human system), suitability for delivery and transportation, etc. The technical system must exhibit other sets of properties in order to ensure the necessary behavior, which include strength, dimensions, form, surface properties, etc.

When we intend to buy a technical system we determine not only what it shall be able to do, we also specify its properties. These properties of a technical system are called *requirements* when formulating a design task and assigning it to the designers.

PROPOSITION 7.1

A technical system must be capable of performing the desired functions to fulfill its requirements, and must also possess certain other properties in an appropriate measure and form.

A technical system possesses certain properties, for instance with respect to minimum life or corrosion resistance, even if the requirements for these properties have not been explicitly stated.

Each technical system possesses all types of property, but their states of embodiment and measures (magnitude and units) can be present in different levels. The sum of these measures determines the total *quality* (both of design and of conformity) of the technical system.

The following sections of this chapter will present various ways of classifying the set of properties, according to various important criteria which can make some of the tasks of the designer easier. Particular criteria are concerned with specifying the characteristic requirements for a TS, assigning a task to the designers, for comparison, evaluation and verification, etc. The importance of a correct and complete task definition for designing a technical system is emphasized at this point. It will initially be in the form of a *requirements specification* issued from the viewpoint of a customer or user, and is transformed by the designer into a *design specification* to guide the design process. It is also important for the designer to comprehend the relationships between the properties, because the nature of designing comprises discovering and establishing suitable design properties.

7.1 Characteristics of Technical Systems As a Class

The class of objects that we call technical systems is distinguished by a set of typical characteristics. The most noticeable of these characteristics is the *diversity* of all elements, which has no analogies to any other field. As an example, the nomenclature in mechanical engineering includes many thousands of kinds of technical systems, even if we do not consider the variations of types. Apart from the diversity and multiplicity of functions that are reflected in this nomenclature, there is also a numerous set of other properties.

The *behavior* of a technical system is always directed towards executing a set of transformations, even though its behavior is very different from that of another TS. The behavior of any technical system is closely related to its structure (which was discussed in chapter 5), because the elements and relationships that exist in the structure determine the properties and behavior. Behavior is *determinate* and *controllable* (it can be adequately predicted, and can be regulated by suitable control inputs) under normal circumstances, except if the system is damaged in any way.

PROPOSITION 7.2
The behavior of a technical system is determined by the structure of that system.

PROPOSITION 7.3
A relatively closed system with a given structure exhibits a single behavior, determined by its structure.

PROPOSITION 7.4
The observed behavior does not uniquely determine the structure that caused it. The same behavior can be realized by a number of different structures.

Many important technical systems are *complicated* or *very complicated*, and consist of thousands and even tens of thousands of parts (i.e. smaller technical systems). A natural result is the high cost of these technical systems. This is particularly true for those technical systems that are produced in very small quantities (especially one-off items) as capital equipment, since they are generally complicated and expensive.

A further noteworthy feature of technical systems is their wide *distribution* and application in all fields of the economy and society, as figure 6.1 illustrates.

At present, the *basic materials* of construction for technical systems are still metallic, mainly steel, but the use of natural and artificial non-metals continues to increase, particularly plastics and other polymers, composite materials, and ceramics.

7.2 Categories of Properties of Technical Systems

The set of properties of technical systems can be classified in many ways, according to various criteria and frames of reference. The following sub-sections serve to discuss the most important of these categories. Each property can be usually classified according to each criterion, but the boundaries between sub-classes will usually not be strictly defined, and may be subject to individual interpretation.

7.2.1 Categories of Properties by Way of Observing

(1) external properties
(2) internal properties

Some properties of technical systems are easily observed, either by human sense organs, or by various assisting devices (e.g. measuring or sensing systems). We can term these the *external properties.* They comprise form, dimension, color, and many others. A second grouping of properties are so difficult to observe that an expert is needed to determine their existence and measure. These we will term the *internal properties.* The boundary between these sub-categories is not well defined. The strain on the surface of a mechanical component may be regarded as an internal property (measurable by an expert), or as an external property (displayed on a meter). The actual strength of that component is almost definitely an internal property, even an expert can only determine the strength of a particular component by destructive testing.

It would be inaccurate and relative if we only distinguished the properties according to this viewpoint. A more precise definition must be based on the principles of systems. In this view, the external properties are the relationships of the technical system to its environment. The internal properties consist of (a) the relationships between the elements of the system, and (b) the properties of those elements. The strain on a mechanical component is thus an internal property, its transduced measure as displayed on the indicating instrument is an external property.

The relationships at the lower levels of discrimination are usually not considered. They reflect the properties of the elements of a TS that are given by their internal structure, such as material or electrical properties. Such properties should only be considered in those special cases where they are important for the functions of the technical system (e.g. the fatigue resistance of a component, which is related to reliability).

The user (or the direct operator) of a technical system is primarily interested in the external properties.

7.2.2 Categories of Properties by Causal Relationships

(1) primary properties – the causes
(2) secondary properties – the consequences

Causal relationships exist between the properties. Causality is an important type of relationship, as discussed in section 3.1, 5.1 and appendix B. Being able to distinguish causes from consequences *post hoc* (after the event) must rank among the basic abilities of every engineering designer. Designing as a process takes the opposite direction – for the desired effects (as consequences), the causes must be selected and established in such a way that they will precede (and thus cause) the desired consequences when they operate.

Examples of such causal relationships are:

- the high temperature of the rapidly compressed gas (air) in the cylinder head of a Diesel motor (the cause) ignites the injected fuel (the consequence);
- a lack of stiffness of a lathe bed (the cause) results in inaccuracies when machining a workpiece.

The sequence of a cause occurring in time before its consequence is the necessary condition for the time that elapses in all processes.

7.2.3 Categories of Properties by Functional Dependence

(1) dependent variable properties
(2) independent variable properties

The natural sciences present many theoretical and empirical relationships (mostly in the form of equations) that show how a property depends on others. Examples are:

(a) the relationship between hydrostatic pressure in a liquid and distance below the surface level;
(b) the relationship between dimensions, form (shape and surface finish), material (strength capability), temperature, etc., and the capacity of a part to carry a load;
(c) the relationship between flow velocity and pressure loss, and pipe diameter and roughness;
(d) the relationship between the output of a fan and the geometry of its blades;
(e) the relationship between internal pressure and compressive buckling of the toroidal transition to the toro-spherical head of a thin pressure vessel [85].

All of these properties are related by mathematical functions.

Those properties of a TS that are influenced by other properties are termed *dependent variables* (e.g. load-carrying capacity). Those that influence another property are termed *independent variables* (e.g. material, form, dimensions, tolerances, surface finish).

These terms are also to some extent relative, and the definition of which variable is "dependent" can change with the application. For the process of problem-solving, any equation (or inequality) represents and models a conceptually invariant property of a technical process or system. Properties whose measure is known for the purpose are termed "independent", those measures that are unknown are declared "dependent". The equation (or inequality) can generally be re-written to show a different variable as "dependent". Some equations can be solved explicitly, some require iterative or approximate solutions.

How some of these categories of properties are interrelated may now be demonstrated. Some external properties of a technical system are under the direct control of an operator - they are dependent or secondary properties in that role. The same properties may be independent or primary with respect to other parts of the TS.

7.2.4 Categories of Properties by Possibility of Quantification

(1) properties that are easily quantified
(2) properties that are difficult to quantify
(3) properties that are not quantifiable

Direct measurement to yield repeatable measures is possible on properties that are easily quantified. Where the properties can not be directly quantified, a point-rating scheme can be used. This results in a ranking or grouping of the measures of a property into classes according to certain criteria, with coarser or finer steps. Many problems (e.g. of accuracy and repeatability) arise through such a scheme, connected both with the person performing the assessment, and with the difficulties of establishing the numerical evaluations. This method is certainly more accurate than a purely subjective evaluation from "feelings", which may produce simplistic results such as "good" or "bad". This problem will be discussed in more detail in chapter 8.

The introduction to this chapter stated some examples of properties in these three categories.

7.2.5 Categories of Properties by Importance

(1) very important and essential properties, e.g. function, safety
(2) important properties, e.g. reliability, life capability, price
(3) less important properties, e.g. packability, storage shelf-life
(4) relatively unimportant properties, e.g. appearance, color

The various properties of a technical system have a range of importance. We must also be aware of the relative nature of the importance of a property and its assessment. Importance depends on the nature of the product, the circumstances of its use, the market, and many other factors. Sometimes rapid delivery is essential, in other cases easy dis-assembly of the technical system is most important. The steps of importance can vary widely. The relative importance of a particular property plays a large part in evaluating a technical system, and is frequently stated in terms of a "weighting factor". With this technique (explained further in chapter 8), an overall evaluation of a TS can be influenced both by the rating that is achieved with respect to a property (see 7.2.4), and the importance of that property.

7.2.6 Categories of Properties by Technical-Scientific Areas

(1) geometric properties, e.g. width, height, angle, center-distance, shape, symmetry
(2) kinematic properties, e.g speed, acceleration
(3) mechanical properties, e.g. strength, elasticity, deformation, density
(4) thermal properties, e.g. conductivity, thermal capacity, radiative emissivity
(5) electrical and magnetic properties, e.g. capacitance, voltage, resistance, conductivity
(6) optical properties, e.g. focal distance, refraction, reflection, polarization
(7) acoustic properties, e.g. damping, echo, noise, frequency response
(8) chemical properties, e.g. aggressiveness, concentration, corrosion, affinity, fugacity

Classification of properties according to the technical and scientific areas corresponds to the general structure of education and course contents. Listing these properties in this form produces an interesting picture of the breadth and scope of the properties of technical systems.

Interlude
Before we consider further groups of properties, it is worth investigating whether the classes of properties up to this point have contributed to the goal of helping engineering designers in their work. The most important aspect of the preceding classifications is to explain the relationships between the TS properties, which is achieved especially in the first three groups. The survey of properties according to technical-scientific areas of knowledge serves to highlight the variety of the properties of technical systems.

Naturally, none of the previous classifications is sufficient to completely specify all the properties that the designer needs to consider, especially in connection with systematic design work. Design aids such as the categories of properties described in this chapter should give complete coverage for all types of technical systems, from component parts to complex plant. They must also take into consideration the technical, economic, ergonomic, societal, transport-oriented, legal and other aspects. Providing information in this form is a complicated task that has not produced satisfactory results to date.

Extensive collections of properties of all kinds have been published by Kesselring [132] and VDI [18]. They list ten basic groups of properties:

(1) countable properties (delivering integer numbers)
(2) geometric properties
(3) mechanical properties
(4) thermal properties
(5) electrical and magnetic properties
(6) optical properties
(7) acoustic properties
(8) chemical properties
(9) production and assembly properties
(10) operational properties.

Abbreviation	Class of Properties	Questions about Class	Groups or Examples of Property Class
Fu (1) Ef	Function Effect	What does the TS do ? What capability does the TS have ?	Working function Auxiliary function Propelling function Controlling & Regulating function Connecting function
FD (2)	Functionally Determined Properties	What conditions are characteristic of the function ?	Power Speed Size Functional dimensions Load capacity
Opp (3)	Operational Properties	How suitable is the TS for the working process (operation) ?	Operational safety Reliability Life Energy consumption Space occupation Maintainability
Erg (4)	Ergonomic Properties	How is it to be operated, and what influence does the TS have on human beings ?	Operator safety Way of operating Types of secondary outputs Requirements for human attention
Aes (5)	Aesthetic Properties	What influence does the TS have on human sensory feelings ?	Form Colour Surface distribution Surface juxtaposition
Di (6)	Distribution Properties	How suitable is the TS for transport, storage, packing ?	Transportability Storage suitability Packaging suitability Suitability for commissioning
DP (7)	Delivery and Planning Properties	When can the TS be delivered ? Manufacturing quantity	Delivery capability Quantity production One-off production
LC (8)	Law Conformance Properties	Does the TS conform to laws, codes of practice, standards ?	Standardized Patent clearance Legal
Mfg (9)	Manufacturing Properties	How suitable is the TS for manufacture ?	Manufacturability Assemblability Manufacturing quality
Des (10)	Design Properties	With what are the external properties ?	Structure Form, Shape Dimensions Materials Surface quality Tolerances Type of manufacture
Ec (11)	Economic Properties	How economic is the working and manufacturing process ?	Operating costs Manufacturing costs Effectiveness Price Manufacturer
Liq (12)	Liquidation Properties	How easy is the TS to liquidate ?	Re-cycling Danger of wastes

Fig. 7.1 Classes of Properties of Technical Systems

Groups 2 to 8 correspond to areas of technical-scientific knowledge (see 7.2.6), whereas groups 1, 9 and 10 consider topics and properties not covered by the classifications discussed in the earlier sub-sections. These properties do not cover all possibilities, they nevertheless show the wide range of properties of technical systems.

7.2.7 Categories of Properties by Needs for Design Work

A survey of these categories is shown in figure 7.1. The questions listed in this figure should help to identify the contents of each category, the examples are intended to assist understanding, and this sub-section serves to clarify the categories.

7.2.7.1 Effects (Ef), Functions (Fu).
According to Statement 4.5, the working effects and other accompanying effects necessary to accomplish the transformation are realized by the execution system comprised of humans, technical systems and active environment in the range of activities of engineering. The technical system must therefore have the necessary functions (i. e. capabilities) to achieve the desired effects.

In modern usage in the technically developed regions of the world, technical systems already supply most of the necessary effects for the majority of transformations. Humans reserve particularly some of the regulating and controlling effects for themselves.

The most important class of properties of each technical system comprises the functions. Even if the other properties of a TS are extremely well developed, if the functions are not fulfilled (the car does not travel, the lathe's spindle does not rotate) the technical system is useless for the purpose of driving a desired transformation. It could still fulfill a different purpose, for instance in the social context as a decoration, an exhibit, an educational or financial investment object. Some technical systems that have reached the end of their useful life (see sub-section 7.2.7.3), and can no longer fulfill their original purpose (and functions) are also used for such social purposes.

It is worth remembering that the term *function* in this book is used for that property of a technical system which makes it able to do something. This property, for designing the TS, is abstracted from certain conditions of a given situation. For example, all types of motor vehicle, aircraft, ships, conveyors, cranes, lifts have in common the function of transporting - the operands, and the conditions vary. The function of transporting with ground-contact and without rail guidance is achieved by motor vehicles. The same function, but limited to the operand "human being" is achieved by passenger vehicles (cars and busses). The same function for speeds over 200 km/hr is the property of racing cars; and so on. Figure 7.2 illustrates the different degrees of abstraction of the function "transporting" according to the number of imposed conditions fulfilled by that function. According to this example one can conclude that the *behavior* of a technical system is equivalent to (and a limiting case of) the function that includes all conditions of the concrete situation. We are therefore in these examples no longer dealing with a broad range of technical systems that show a certain behavior, but

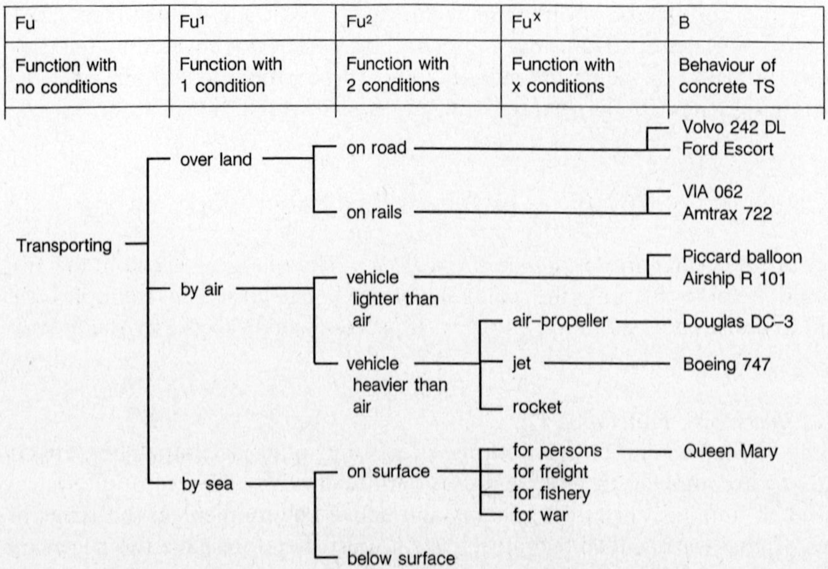

Fig. 7.2 Examples of Functions at Various Levels of Abstraction

rather with a class that shows variations in the secondary properties (see Proposition 7.2). The importance of this insight will become clearer when considering the general methodology and particular methods of design work.

Functions (and behaviors) exhibit different levels of complication. Consider some of the functions needed to drive the process "manufacturing paper": drying, blowing hot air, connecting a hub to a shaft for transmitting torque. All of these partial functions (functions that are part of the higher-level function) occur at a fairly high level of abstraction, and some of the partial functions are more complicated than others. Connecting a hub to a shaft is an elementary function, the degree of complication increases progressively until the complete function of manufacturing paper is reached. The example thus shows a hierarchy of functions. "Manufacturing paper" needs a series of functions such as "drying", which needs a series of functions including "blowing hot air", which needs a series of functions including "increasing pressure", which needs a series of functions including "connecting an impeller hub to a drive shaft", and so on. At each step, the function is realized by a series of partial functions that each depend on the technology employed to fulfill that function.

PROPOSITION 7.5

Every function (or behavior) is a system that contains as its elements a set of partial functions (or partial behaviors), and that is decomposable to elementary functions (or elementary behaviors). At each level of complication (from total function to elementary function) we can work with various degrees of abstraction (ranging from an abstract function to the actual behavior of a concrete technical system).

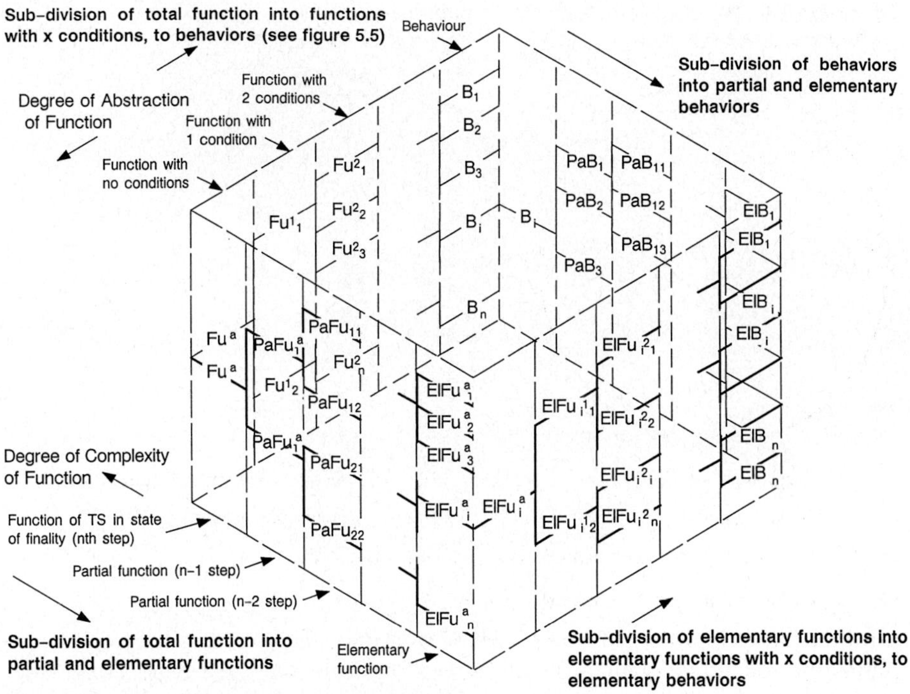

Fig. 7.3 Function Diagram: Decomposition of Functions

The particular functions "connecting" and "separating" are assumed to be elementary functions at a high level of abstraction. They combine a series of functions from various steps in the hierarchy of abstractions, and various stages of imposed conditions (see figure 7.3, and compare figure 5.5)

Because various structures can exhibit similar behavior, each function can be realized by a number of different technical systems, according to Proposition 7.4. The degree of complexity of a technical system, from the most complex plant to the simplest machine component, corresponds to the level of complication of the function that it realizes.

Conversely, any one technical system can perform a variety of functions. A screw can, for instance, connect two parts together, or transform a rotary motion into a linear motion, or form an adjustable stand-off spacer between two parts. Some technical systems are deliberately designed in such a way that they can perform as many functions as possible (within the existing economical, financial and physical constraints). This can be achieved by aggregating some functions. The mutual relationship between the set of functions and the set of technical systems at a certain level of abstraction is shown in figure 7.4. The possibility that a technical system can fulfill a larger number of functions generally reduces as the degree of complexity of these functions increases.

As stated earlier (statement 4.5), a working effect can only be achieved if a number of additional effects are also realized. Consequently, the category of functions

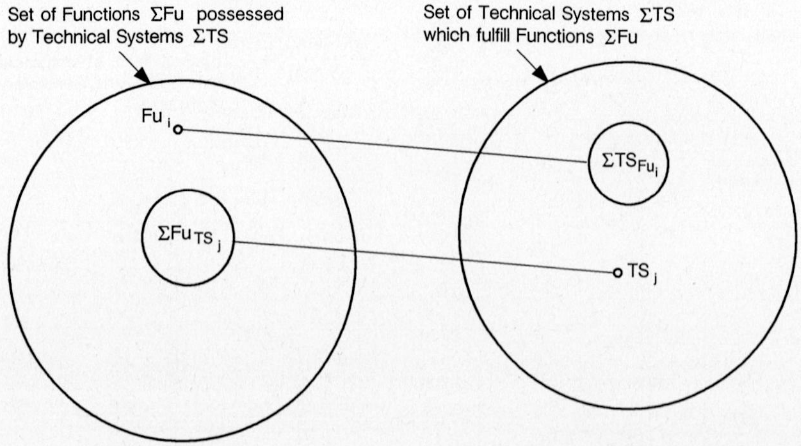

Fu_i ... an arbitrary Function

TS_{Fu_i} ... TS that fulfills Fu_i

TS_j ... an arbitrary Technical System

Fu_{TS_j} ... Function possessed by TS_j

Fig. 7.4 Relationships between Functions and Technical Systems

can be divided into the sub-categories of working functions, auxiliary functions, propelling functions, regulating and controlling functions, and connecting functions (analogous to the processes, see section 4.1.1). Some of the examples in chapter 4 served to illustrate and explain these terms (particularly in sections 4.1.1 and 4.3). We must acknowledge that even though these sub-categories of functions are relative, they refer to all levels of complication of a technical system, including the TS in its state of finality, its partial systems, constructional groups, sub-assemblies and component parts. Usually an auxiliary function (or any other additional function) of the main system is performed by a partial system. The auxiliary function of the main system then becomes the working function of the partial system. For instance, in a lathe, the auxiliary function of holding a workpiece can be realized by a three-jaw self-centering chuck. This is a partial system (or sub-system) for which "holding" is the primary working function, and "attaching to a lathe spindle" is an auxiliary function for the chuck. Differentiating the functions in this way helps to clarify the task given to a designer, and coordinate the efforts of individuals within a design team.

7.2.7.2 Properties Determined by Effects and Functions (FD).

The example in the previous section demonstrated how a function could generally be progressively concretized by stating additional conditions, e.g that transporting should be performed on humans, over land, and at a certain speed. These are three of the *conditions* that restrict the solution space, but they still permit a wide range of possible solutions. To transport humans over land, a car, a bus or a bicycle would be permissible means to fulfill the primary function as stated. Therefore, in order to realize a concrete technical system, additional conditions and characteristics must be specified. In the example, these conditions could be the number of persons to be transported, the locations of start and finish, the distance, the quality of surface (road), and so on. All of these statements represent further conditions on the function of transporting. A *constraint* is a particular form of condition that sets strict limits to acceptable performance and other criteria.

Similarly, other functions can be concretized (and the solution space restricted) by specifying further characteristics of the TS, a task usually performed by an expert.

On the basis of this analysis it can be concluded that the added items of information about the transformation constitute the *functionally determined properties*, the additional properties which are determined by considering constraints and conditions on the functions. These include details about the operand, the technology, the conditions under which the transformation is to happen, and so on. These important properties are closely connected to the function or the behavior of the technical system. Such properties are always stated in manufacturer's catalogs and advertising literature, because only through this information does the purpose and possible use of a TS become obvious to the potential user, and to the designer who wishes to incorporate that TS into a system. These items of information are termed the *parameters*, especially if they contain numerical data.

From the point of view of design method, the functionally determined properties act as a screen. By specifying additional conditions we restrict the variety and number of available solutions, and the set of means for performing the desired transformation, until the limited selection that remains contains only those systems that can fulfill all the stated conditions (including requirements and constraints). The remaining proposed systems are then evaluated and optimized with respect to suitable properties.

7.2.7.3 Operational Properties of the Technical System (Op).

A required transformation is achieved by the effects exerted (on the operand) by a technical system while it operates. How suitable a technical system is for such operation is characterized by a group of *operational properties.* These include reliability, safety, life, energy consumption, consumption of auxiliary materials, space requirement, suitability for maintenance, operability, adjustability, replaceability of parts, acquisition (first) costs, operational (running) costs, and so on. In many applications of TS where the functions can easily be realized, the operational properties are becoming more important. The requirements with respect to operational properties change over a period of time, and are different according to the nature of each field of engineering.

The requirements for reliability and safety are especially high in those cases where the human is an operand in the process, e.g. in transport. Most people are reluctant to place a price on the life of a human. Opinions about a suitable life-span for technical systems are changing; with the rapid development of engineering the edict that "a part or item of equipment must have a long life" is no longer generally valid. In many cases a shorter life-span is required, in others an extremely long life is desirable. The ways in which the life of a technical system may end are varied, and may be summarized as follows:

- wear-out, when it can no longer fulfill its functions because of general wear and deterioration, or when repair is no longer feasible, the most usual criterion for retiring a technical system in the past;

- catastrophic failure, when an event occurs that renders the system unusable by physical damage;
- obsolescence, when a newer and better system becomes physically or economically available to fulfill the same or increased functions;
- operation of a one-shot device, when a system is intended to be usable only once, and is used, for instance a disposable fire extinguisher or a gun cartridge;
- replacement before completed life expectancy, when a part is taken out of service before it is likely to fail, to avoid catastrophic failure of a larger system, e.g. the turbine blades of an aircraft gas turbine.

Different tendencies are visible for space and weight requirements. Available space and permissible weight are generally reduced, which may also result in reduced costs of acquisition and operation. Size reduction of technical systems is thus a design requirement that must be considered.

While a technical system is operating it generally emits noise, heat, vibrations, pollutants for the environment, and damaging waste products that must be removed. These are collectively termed secondary outputs (see section 4.5). Increased usage of technical systems has resulted in some of these secondary outputs reaching dangerous or even catastrophic proportions, with consequent hazards for human life and the environment. To date the consequences of these properties of technical systems have frequently been ignored, or neglected for economic reasons. If humanity wishes to outlive technical systems, it is vital that the requirements with respect to secondary outputs and environmental protection are appropriately considered in designing the processes and systems.

Measuring the operational properties is very difficult in many cases, particularly for technical systems that are still in the conceptual, lay-out and detail design phases. Some of these properties are only noticeable after longer periods of operation. It is therefore important that designers follow and observe a technical system over a sufficient operational time, and analyze the causes for unsatisfactory operational properties [77]. This is the only source of information that can deliver concrete insights about operational properties.

7.2.7.4 Ergonomic Properties (Erg).

Transformations are realized by the effects exerted (on the operands) by the system of "human being - technical system - active environment", according to statement 3.8. The whole technical result depends on good cooperation between these partners. Society is also generally deeply concerned that this cooperation will not result in damage to humans (with noteworthy exceptions such as war operations). In the past few years an independent scientific discipline has become concerned with studying relationships between humans and technical systems, especially direct interactions between them - *ergonomics* . The practical field of application is known as "human engineering" [21, 71, 131, 134, 209] which is based on anthropology, physiology, psychology, biology, sociology, physics and cybernetics. It includes studies of time and motion, sensory perception, work loading, human data processing and decision-making, affective loading (human stress and feelings), extreme physical loadings (time, force, temperature, pressure, acceleration, radiation, subsistence), loss of orientation (acoustic, optical and tactile), adaptation and orientation to a new reference system.

Ergonomics is not only concerned with safety and industrial illnesses, but also attempts to find the boundaries of human capabilities in extreme situations. It is true that certain of the questions asked in this context are intended for use within a limited field of engineering endeavors, for instance rocketry and space travel. Nevertheless, there are many areas of general application for ergonomic data and results of experiments, for instance laying out the regulating and controlling devices to enable a human being with average capabilities to act and react properly and predictably, usually as quickly as possible, and with minimum physical and mental fatigue, even in an emergency.

The designer's task within the system "human being - technical system - active environment" includes allocating appropriate tasks to the human being and to the technical system. Designers must therefore be aware of the advantages and drawbacks of each of these operators. The relevant factors are summarized in figure 7.5, and their relationships illustrated in figure 7.6.

Concrete data available to designers from studies of ergonomics are:

- capabilities of various parts of the human body
- dimensions of humans (anthropometry)
- capabilities of human beings in various environments
- arrangement of measuring devices and regulating organs with respect to optimal performance of the system "human being - technical system - active environment".

Many special situations exist for which appropriate recommendations and data can only be obtained by controlled experiments [134].

Human	Technical System
- is capable of reaching important decisions based on limited information	- works only according to given orders
- correct reaction even under unexpected conditions	- unexpected (by the human) conditions can lead to disaster
- can perform certain operations in various ways, important when the mechanism is damaged	- number of operations performable by a TS is limited
- is flexible in work programming	- program changes are usually difficult and costly (exception : computers)
- capacity of information reception is limited; reception rate for information is limited	- capacity of information channels can be readily increased
- limited capability of (receptor) sensing organs	- possibilities of increasing sensing parameters
- power and attention reduce with time	- almost constant power and precision
- working capability requires certain conditions : temperature, humidity, pressure, noise	- can be made for any environmental conditions
- thinking operations relatively slow; large probability of errors	- rapid execution of logical operations almost without error
- limited memory and recall capabilities	- almost unlimited information storage; limited by ready access to stored information
- information can be complex; can readily combine and interpret	- form of information basically simple; needs full instructions to combine or interpret information

Fig. 7.5 Characteristic Properties of Humans and Technical Systems

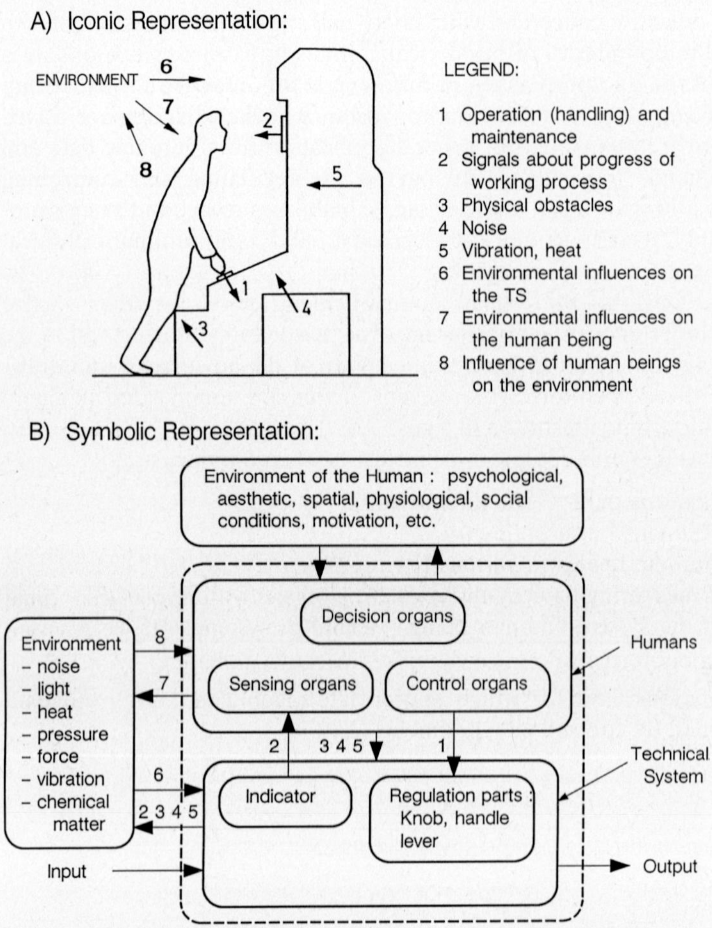

Fig. 7.6 Relationships in the System: Human - Technical System

7.2.7.5 Appearance and Aesthetic Properties (Aes).

The division into ergonomics and aesthetics may seem arbitrary. Both are based on interactions between humans and technical systems, but with respect to ergonomics the human is regarded as a cognitively and affectively active participant. Aesthetics, the study of appeal to the senses, deals particularly with the appearance of a technical system and how humans react to it, mainly as passive observers, although some affective activities can occur.

The tendency to coordinate the useful properties of a technical system with the aesthetic properties goes back to the start of engineering, and is already noticable in crafts and handwork. This principle can also be found in the study of engineering design, Reuleaux [179] included an essay on style in the construction of machines. The forms of older machines are no longer considered satisfactory for our modern sense of beauty.

Later works on designing propounded the rule that any technical system that is purposeful is also beautiful – "form follows function". Only in recent years has a new discipline emerged to study the relationships between the sense of beauty and the rules for establishing the forms of systems. We are not only concerned with the sense of beauty and satisfaction as a result of an aesthetic experience, but also with some more practical results of these feelings. The influence of aesthetic impressions on mental and physical activities has been demonstrated, for instance by showing higher performance with less fatigue of human organisms in suitable aesthetic surroundings. In order to influence the state of an organism, the senses must also be suitably stimulated, and this circumstance provides a connection between the aesthetic and ergonomic properties.

Guidelines for designers are summarized in rules of geometry and metrics of forms which should impart order and organizational value to the technical system [36, 157, 201].

Basic elements of form include:

- symmetry – absolute, relative, contrasting;
- rhythm – purposeful variation of bodies, surfaces, colors, structures;
- structuring – horizontal or vertical directions;
- proportionality – mutual dimensions and relationships between humans and surroundings;
- contrast – of materials, forms, dimensions, colors, textures, light;
- composition – forms, lines, colors.

The formative elements used to achieve the desired impressions, moods and reactions such as peacefulness, attention, pleasure, harmony, order, etc. are:

- form – soft, hard
- material – different impressions with wood or metal
- material structure – natural, prepared
- size or volume
- line – horizontal, vertical, sloping, straight, curved
- color – tone, surface texture: dull, glossy
- text and symbols – on the product or a label
- light and shade

Consideration of aesthetic insights can substantially increase the value of a technical system, compare figure 7.7. This tendency, while principally correct, should

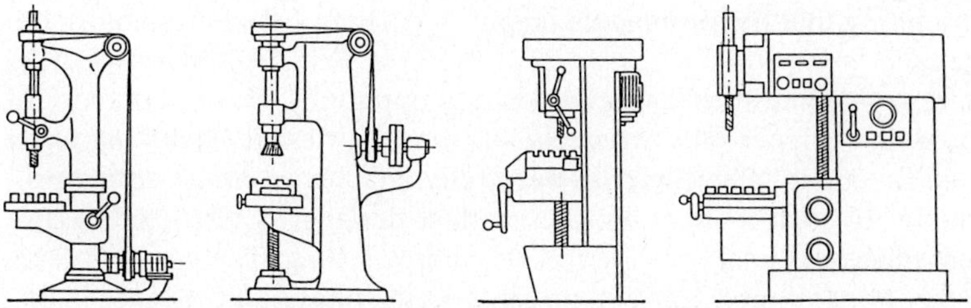

Fig. 7.7 Forms of a Drilling Machine

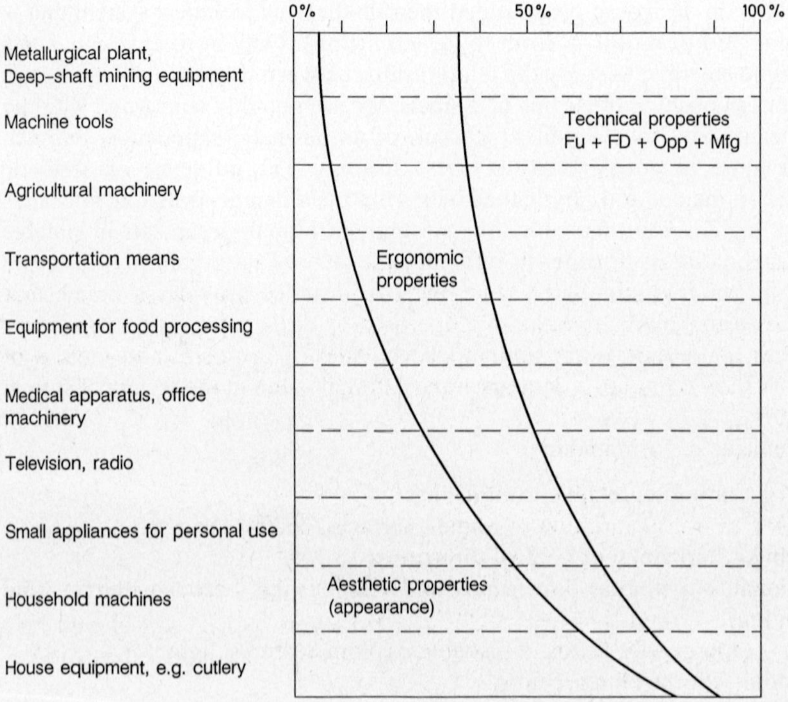

Fig. 7.8 Proportions of Important Groups of Properties for Some Classes of Technical Systems

not be exaggerated, especially if other desirable functions of the system were to receive too little attention. Each field of engineering has its balance between technical, ergonomic and aesthetic properties, as shown in figure 7.8. This group of properties, and the current thinking on appropriate form, are influenced by the state of the art, and tendencies of fashion and style, subjects that are also touched in chapter 11.

This raises the question whether engineers, when designing a technical system, can and should perform the tasks of aesthetic design by themselves, or whether they need the help of a specialist. Engineering designers must at least be aware of the rules and principles of aesthetics, even if a specialist is consulted.

Specialist product designers, known as industrial designers in Britain, are available for these tasks. They concern themselves with aesthetics and ergonomics, especially considering that the problem is one of interaction between humans and technical systems.

The timing for consulting such a specialist is also important. The bad practice of involving a product designer only when the engineering design (including exact placement of all functional elements) has been fully established in the design process is deplorable. In such a situation, the product designer is restricted to aesthetic treatment of the external appearance, i.e. simply styling. The correct procedure is to ask for cooperation as early as possible, in the phases of conceptualizing a solution, or at the latest before the dimensional lay-out is finalized.

7.2.7.6 Distributional Properties (Di).

The technical system is generally manufactured in a factory. A product is usually subjected to a long and arduous procedure leading up to the time of first commissioning for operation. Especially a large product has to be prepared for moving, brought out of the factory space without needing to remove the factory doors, hoisted by a crane of adequate lifting capacity, stored without taking up too much space, transported to the user (by truck, rail or ship, exposed to weather, shocks, etc.), brought into the operational area, assembled in as short a time as possible, and commissioned for operation. For household appliances, the operations of shipping, storing, transporting to the user, and installing are probably the most arduous and hazardous. Distribution makes severe demands on a technical system, which must consequently possess the necessary properties by means of which it is capable of withstanding the process of distributing at minimal cost.

Generally the costs of storage, transportation, and packaging are calculated by volume and weight, and they are directly charged to the product.

Some of the properties that influence the capabilities of a technical system to withstand storage, transport and packaging are summarized together with appropriate design recommendations as follows:

- overall dimensions of the product influence the shipping volume, as the most important parameter for distribution; the maximum permissible dimensions and profiles for transportation of goods should not be exceeded (ruling profiles for rail transport, clearance height of road bridges, etc.);
- form, for instance the shape of protruding parts, can influence the circumscribed volume significantly, which in turn affect the storage, transportation and packaging costs; possibilities of partial dis-assembly should be considered;
- weight must be limited to within the carrying capacity of the means of transporting and lifting; a transport-oriented division of the device into smaller units should be arranged if necessary;
- sensitive and delicate parts that could be damaged in transit; these should be designed to be dis-assembled so that they can be packed separately;
- material may need special treatment, especially if it could be damaged by climatic conditions (tropical, desert, arctic, sea transport etc.); easy replacement may be necessary;
- strength must be sufficient to accommodate the loads applied during lifting and moving; support by additional beams or struts may be necessary, plus provision of adequate hooks, lift eyes, fork-lift slots, towing points, etc., which will ensure that these loads are applied mainly at the anticipated points on the constructional elements that perform the connecting and supporting functions (frame, structural members, bracing, etc.).

Suitability for storage together with reduction of volume for transport can influence the form of a quantity-produced product. Bottles and other containers, and pressed forms such as those shown in figure 7.9, provide examples of such influences.

Some of the requirements for assembly or dis-assembly have been mentioned in this section, and demonstrate another set of properties of technical systems, the manufacturing properties, which are discussed in subsection 7.2.7.9.

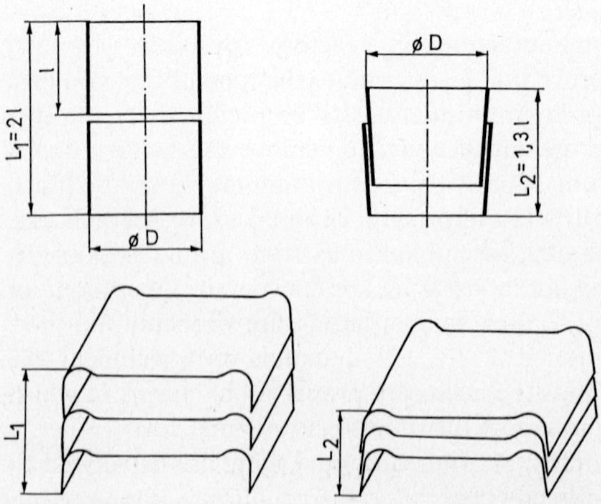

Fig. 7.9 The Form of the Product and its Influence on Storage Volume

7.2.7.7 Delivery and Planning Properties (DP).
For every buyer or customer of a product, the delivery deadline is an important property that can frequently be decisive in awarding a contract for capital goods. This delivery deadline is also important for the designers because it establishes the deadlines for their work.

The designer's work is also influenced by the quantity of the technical system to be manufactured. This factor influences the level of detail in working out various design documents, and the progression of the design work, as discussed in chapter 10 and 12. Two limiting cases can be summarized as follows:

- A single technical system, or a small number, is to be produced. The problems reside mainly in achieving the functions and desired properties, especially by careful prior thought and work, because it is usually not possible to build and test a prototype. Work in the conceptual and preliminary lay-out phases of designing must usually be very thorough. In this case, the amount of care expended on formalizing the detail drawings and design and manufacturing documents will be less than in the second case.
- When producing a larger number of nominally identical technical systems, in large batch or mass production, the drawings and design documents must be more carefully formalized. It is usual to manufacture and test a prototype, to verify that the functions have been achieved, and to engage in development work to bring any deficient properties up to the desired standards, and to reduce costs. Special tools, jigs and fixtures for manufacture must also be prepared.

The delivery and planning properties of a technical system, namely whether it is made as a one-off, small batch, large batch or mass-produced item, have certain consequences for the quality of the system, and achievement of individual properties.

7.2.7.8 Law Conformance Properties (LC).
Every country, state, province, city, or other community issues laws, bye-laws, codes of practice, regulations, etc. These are usually intended to provide some measure of protection to the customers and users of technical systems, and in some cases also to the manufacturers. How the technical system is viewed by the law, and how it complies with the regulations, is an important property.

Regulations and standards have different force under the law. Many are compulsory, and laid down by the legislative procedures. Some are enforceable by indirect application of laws, for instance through the laws of contract and tort. Other regulations rest on voluntary agreements between corporate bodies, resulting in the international, national, regional and technical society standards. A further group is formulated as an agreement of a group of involved people, such as the in-house standards of various companies. Even such information as climatic or political conditions could be included in this discussion. They also include the codes of ethics and other rules laid down by the governing bodies of the relevant profession, particularly the engineering institutes and associations.

Designers must know the regulations and standards that are relevant to the technical system to be designed, including those originating in the country or region of manufacture, and those applicable to the regions in which the technical system is to be used. This point often receives too little attention, and may result in the goods being refused on delivery, returned at a border crossing, or cause extensive procedures in courts of criminal or civil law. Severe penalties can result, particularly if safety regulations have been overlooked, and human life has been endangered or lost.

Similarly, designers must also know the patent laws and rules of the countries of origin and of use of the technical system. Search by a competent patent engineer or lawyer may be needed to ensure that no patent rights are violated. It is irrelevant whether designers have found the idea for themselves, an agreement to permit use must be reached if the idea was previously patented.

Laws concerning the responsibilities of manufacturers, vendors, engineers, etc. differ from country to country. Typical law demands that a product will be manufactured (a) according to the accepted rules of engineering and science, (b) to comply with health and safety regulations, and (c) to comply with accident prevention laws.

Interlude.
The properties discussed so far are classified as the external properties of a system, those that primarily interest the user of a technical system. The main requirements from the user are that the system will fulfill its functions, and possess the necessary parameters, appearance, ergonomic properties, etc. All of these properties also affect the designer and manufacturer, but particularly the properties related to transportation, delivery and law conformance. The next sub-sections deal with categories of properties that are of less concern to the user, but are of direct interest to the manufacturer and the designer.

7.2.7.9 Manufacturing Properties of the Technical System (Mfg).
These properties show whether a technical system is suited to manufacture and assembly, or whether it can be made at all.

How many manufacturing operations are needed to achieve the specified form of the component? How difficult are these operations? What preparation for manufacture is needed? Which special manufacturing and measuring tools, jigs, fixtures, dies, or instruments need to be designed and made? What manufacturing accuracy and precision is demanded? What has to be adjusted or fitted during assembly? What must be tested? What must be run in? Can the technical system be manufactured on the existing equipment, must a new plant, system, machine or tool be purchased or manufactured? All of these questions are of prime interest to the manufacturer, and to a large extent they govern the manufacturing costs.

These questions are also of prime interest to design engineers. The choices and decisions made by designers are responsible for about 60 to 80% of the total cost of a technical system (compare also section 7.6). The engineer must therefore have adequate knowledge about any aspects that influence manufacturing and the quality of the manufactures TS. This knowledge should include manufacturing methods and their relative costs, material availability, influence of tooling, tolerances, need for provision of adjustment [101], and a host of other factors. Many of these are discussed in the next sub-section.

The demands regarding manufacturing properties change according to the type of manufacture (one-off, batch, series, or mass production), and with developments in the working processes and means in manufacturing engineering (the state of the art in this field). Recent examples of such developments include numerically controlled machine tools, robotics, and flexible manufacturing cells. A transition to automated manufacture, and mechanized or automated assembly results in requirements for qualitative or quantitative differences in the properties of the parts and technical systems.

Delivery deadlines also have some influence on the choice of solution for a technical system which is to be designed. Some manufacturing processes and machine lay-out arrangements have longer transit times than others, and need different documentation for recording manufacturing progress. Such considerations will, for instance, influence the choice between casting and welding for housings.

The most important question in this respect concerns the method of manufacture. That this viewpoint is not a constant is demonstrated by the example of welding compared to casting. Some time ago the trend was to replace castings by welded or assembled parts. Further developments of materials and casting techniques have increasingly led to a reversal of this trend.

7.2.7.10 Design Properties (Des).

Users generally treat a technical system as a "black box". From directions and instructions, and from experience, the relationships between inputs (pushing a button, setting a lever) and outputs (the behavior of the system that possesses certain characteristics, properties and parameters) are known to the users. They also know that the system must be lubricated, cleaned, maintained, serviced, periodically tested, etc. following the advice given in the instruction manuals. What causes the behavior and other properties does not in general interest users – they wish to raise steam, cut metal, cool milk, grind coffee, and perform many other tasks, aided by reliable devices that should have good appearance.

Immediately when a user unscrews a cover, the shafts, levers, gear wheels, bearings, springs, circuit boards, integrated circuit packages, connectors, valves and other enclosed elements of the mechanical, electrical, hydraulic or other internal features become visible. The world of the expert opens, to reveal the work of the engineering designers who created the system, and the workers who manufactured and assembled it, and must service and maintain it. Looking into the user's "black box" uncovers a set of other properties of technical systems that are hidden for the layman. These internal properties are termed the *design properties* because they serve as means for the designer to create the desired external properties.

A selection of these design properties may include: thermal efficiency, overhead camshaft motor, two-state (on-off) switching, hardness, transmission ratio, strength, stiffness, dimensions of parts, tolerance, material, heat treatment, 8-bit word length, serial or parallel transmission, and so on. It is obvious from these examples that this set can be ordered into classes, and various interrelationships can be seen. Thus strength depends (frequently in some non-linear way) on dimensions, material and form, whereas hardness depends on material and technological processing (but is also correlated with strength). Thermal efficiency and compression ratio are properties connected with accomplishing the transformation, the conversion of chemical energy in the liquid fuel into mechanical energy at the shaft in an internal combustion motor.

Properties such as the ones listed in the last example can be found only in technical systems that perform (guide and drive) these particular transformations. Other properties such as strength, stiffness, hardness, corrosion resistance, wear, etc. occur in a much wider variety of technical systems. We can therefore classify the design properties into the following broad groups:

- **Specialized design properties** are connected with the transformation to be achieved. It is sometimes difficult to distinguish between the functionally determined properties and the design properties. The group into which they are allocated is less important. The fact that they appear in one group or other is essential.
- **General design properties** are those that occur in most technical systems. These may differ from one broad field of engineering to another. In mechanical engineering they typically contain strength, stiffness, wear, hardness, elasticity, corrosion resistance, etc.
- **Elementary design properties** are the means with which the designer achieves all other properties. They consist of the following:
  - for higher levels of complexity of the TS
    (1) structure
  - for the elementary technical system (e.g. mechanical components)
    (2) form (shape)
    (3) size (dimension)
    (4) material
    (5) surface (texture, quality)
    (6) tolerances (dimensional, geometric)
    (7) method of manufacture

PROPOSITION 7.6

All types of properties are achieved by means of the elementary design properties (structure, form, size, material, surface, tolerances, manufacturing methods)

The elementary design properties also exhibit relationships between each other. These relationships cause the enormous complexity of technical systems which make designing such a difficult task.

The groups of elementary design properties need further discussion, at least in rough outline, because they provide the basis for methodical work in engineering design. They are responsible for the need to perform many of the operations of engineering design, namely establishing various structures (see figure 5.4), establishing form, dimensioning, selecting material, establishing surface quality and texture, tolerancing, and proposing manufacturing method (at least by implication).

7.2.7.10.1 Structure (Anatomy of the Technical System)

As postulated in Proposition 7.2, structure and behavior are the most important properties of a technical system. Structure uniquely determines behavior. Any structure consists of a set of elements (at various levels of abstraction) and the relationships between them. These general observations can now be concretized for a particular class of technical systems, for example from mechanical engineering.

Within the vast variety of technical (machine) systems, the constructional elements represent the systems of lower degree of complexity. The structure of the hierarchy of complexity is illustrated in figure 6.3, and the classification of the structural elements according to complexity is shown in figure 6.2. If we restrict the discussion to the lowest level of discrimination, the machine system is composed of a set of machine parts that constitute the constructional elements of the higher systems. We must be aware that this interpretation corresponds closer to that of elementary functions (see Statement 5.12) than to that of the definition of constructional elements according to figure 6.2. A plain sliding journal bearing is for instance a machine element that may consist of a number of parts (housing, bearing shells, bolts, etc.); yet as a unit, it is a means to fulfill the elementary function "permit rotation, transmit radial force". Naturally it is not difficult to disassemble that bearing, and to designate the individual shell and housing items as machine components in the sense of the definition of parts at the lowest level of complexity.

From the viewpoint of mechanics we can describe the structure as follows: The machine consists of pairs of constructional elements that are connected in kinematic chains. The machine is comprised of kinematic chains.

PROPOSITION 7.7

Every technical system (machine system) can be decomposed into partial systems (machines, constructional groups, sub-assemblies, constructional elements) which in their prescribed combination fulfill the partial and elementary functions of the TS.

The relationships between the systems represent an important means to achieve further or different properties, the latter may be qualitatively or quantitatively different. For instance on a car with rear-wheel propulsion, we do not achieve the same result if the mechanical connection between the motor and the body is achieved by placing the motor in the front or the rear end of the vehicle. In spite of using the same basic elements and scheme of mechanisms, the different spatial relationship of motor placement causes differences in such properties as vehicle handling, noise in the passenger compartment, etc.

PROPOSITION 7.8

The behavior of a technical system depends not only on the sum of behaviors of the elements, but also on the coupling relationships between these elements.

Couplings between system elements can be of different kinds, e.g. mechanical, electrical, chemical, magnetic, time or space couplings, or any useful combination (see also appendix B). A complete classification is not possible within the framework of this book. The important relationships occurring in machine systems (products of mechanical engineering) are used in the following as typical examples to guide the reader.

By means of *mechanical couplings* the movement of a point or body in space can be restricted or constrained. A point has a maximum of three degrees of freedom, a body can have up to six (but see [61]). The following couplings can occur:

- scleronomic (stationary), time-independent, which can generally be represented by a system of independent equations of conditions;
- rheonomic, time-variable, which can be represented by equations in which time is a variable;
- holonomic, which can be represented by relationships between point coordinates and time, i.e. scleronomic and rheonomic couplings;
- non-holonomic, which can be represented by differential equations that are not integrable in closed form.

Limiting the number of degrees of freedom results in (compare also figure 5.10):

- unilateral couplings, in which the motion of a body is only restricted in one sense of motion, for instance contact between two bodies, a body suspended by a string, or a one-way rotational coupling (non-return clutch). This form of coupling can be represented by inequalities analogous to the groups above;
- bilateral coupling, in which the body is compelled to perform an accurately defined motion, or to take up an accurately defined position, for example by a linear or rotary bearing. In practice, a rigid coupling of this type is difficult to achieve, but a combination of two unilateral couplings within a short distance (a mechanical clearance) is easily achievable.

One can further distinguish:

- force couplings, through which a certain force is transmitted;
- force-free couplings, which do not transmit force (kinematic couplings).

This does not imply that certain forces do not occur in the second of these classes of coupling.

Other types of couplings include:

- electrical coupling – conductive connections or isolations;
- thermal coupling – heat transmitting or insulating;
- chemical coupling – aggressive or neutral, acidic or basic, oxidizing or reducing, whereby any aggressive coupling can lead to chemical action and corrosion;
- magnetic coupling – field-exciting or magnetically screening;
- feed-back coupling – feeding a part of the output signal back into the input. This coupling is prevalent in control systems. Connections can be in series, in parallel or combined.

7.2.7.10.2 Form

Form is one of the most important elementary properties of technical systems. We can distinguish various types of form depending on what property determines the conditions. For the special case of constructional elements for machine systems (products of mechanical engineering), the term "shape" is usually used. Examples of the influence of the relevant properties on form can be:

- function – aircraft wing profile, screw thread profile, tooth profile for gears, milling cutter profile;
- ergonomic properties – tool handle, steering wheel, driver seat;
- appearance properties – form (shape) of electric iron, machine tools;
- transport and storage properties – containment of gases or liquids, holes or hooks for suspending or lifting, form of a vessel or pressed parts (figure 7.9);
- manufacturing properties – draft angle or edge rounding and fillets for castings or die forgings, run-out for slot or end milling cutters;
- economic properties – simplest form;
- strength properties – beams of constant stress, fillets in stress concentrations;
- wear properties – adjustable guides;
- material – raw material sizes and forms (solid scantling shapes) from suppliers, liquids or gases in pressure containers.

Even if a form is largely determined by one set of requirements, it must still be investigated from the viewpoints of other properties, especially appearance, manufacturing, strength and economic.

7.2.7.10.3 Dimensions (Size)

Sizes and dimensions influence practically all other properties. They too can be conditional on various properties, we speak of functional dimensions, etc.

A few examples of dimensions determined by properties are as follows:

- function – piston diameter, crank throw (radius), belt pulley diameter;
- operational properties – safety factor on calculated size, corrosion allowance, larger dimensions to permit easier access for maintenance;
- ergonomic properties – dimensions of operation control platform, spatial arrangement of control levers;
- appearance properties – ratio of dimensions of various parts, area distribution, golden section;

- transport properties - maximum dimensions to conform with rail transport limits, guiding use of standard sizes for packing crates and boxes;
- manufacturing properties - minimum manufacturable wall thickness of castings or forgings, minimum bending radius;
- economic properties - smallest possible dimensions to reduce material costs;
- strength properties - dimensioning of shafts, size of bearings, face width and pitch of gear teeth;
- stiffness - dimensions of machine tool bed;
- wear - allowance for wear of guidance surfaces;
- material - according to the dimensions of available raw material forms;
- prescribed space and connections - standardized flanges.

7.2.7.10.4 Material

The material selected by a designer from which an elementary technical system is to be manufactured is conditioned by the external properties. Some examples follow:

- function - thermal insulation, electrical isolation, chemical resistance, vibration damping;
- operational properties - quality material for extended useful life or enhanced reliability and safety;
- ergonomic properties - cushioning of seats in vehicles;
- appearance properties - panelling, polished brass, dull-finish aluminum;
- transport properties - corrosion resistance during ocean transport;
- manufacturing properties - selecting carbon steel with a carbon content of 0.3% instead of 0.2%, for better machinability;
- economic properties - selecting a well-priced material that is sufficient for all requirements instead of two or more materials at higher cost;
- strength properties - using alloyed steels for highly-loaded parts, using heat treatments to adjust strength, ductility and hardness for a given duty, special heat-resistant alloys for blading of gas and steam turbines;
- wear - using abrasion-resistant plastics to line a machine tool guideway;
- hardness - sintered carbide cutting tool tips, ceramic valve seats;
- surface quality - selecting an easily machined material to accommodate requirements for smooth surface;
- corrosion resistance - using steels alloyed with Ni and Cr, or using plastics.

7.2.7.10.5 Surface

The term "surface quality" implies the conformation (geometry, texture, flatness, waviness, roughness, chemical affinity or self-protection, reflectivity or diffusivity, etc.) and color of a surface. In order to achieve the external properties of a technical system, the surface quality is more important than is usually assumed. It influences a number of other properties, and must be considered with respect to their requirements:

- function - surface for guideways, bearing surfaces;
- operational properties - longer service life by improved surface quality;

- ergonomic properties – machined surfaces where humans make contact, rougher hand-grip surface to avoid slippage in handling;
- appearance properties – surface quality is decisive for appearance;
- economic properties – minimum of machined surfaces;
- strength properties – smooth surfaces generally show higher fatigue resistance;
- corrosion properties – smooth surfaces are generally more corrosion resistant;
- wear – machined surfaces generally hold lubricant better;
- tolerances – closer dimensional limits demand better surface quality;
- method of manufacture – a certain manufacturing method can only produce certain ranges of surface roughness.

7.2.7.10.6 Tolerances

The tolerance field influences the spatial relationships between parts. Selection of tolerances is influenced by a number of higher properties, such as:

- function – clearance tolerances between shaft journal and bearing shell, interference tolerances for shrink fit between shaft and hub;
- operational properties – interchangeability of components;
- ergonomic properties – wider tolerances can cause higher noise levels in a machine system;
- manufacturing properties – tolerances of parts established such that re-work fitting during assembly is avoided;
- economic properties – closer tolerances cause higher manufacturing costs;
- wear – when mechanical shocks occur, wide tolerances generally cause more wear;
- method of manufacture – a certain manufacturing method can only produce certain ranges of tolerances;
- dimensions – relationship between dimension and achievable tolerances.

7.2.7.10.7 Method of Manufacture

The method of manufacture has an indirect influence on the external properties, by way of the design properties. The following examples show that the relationship between method of manufacture and the elementary design properties is very close:

- strength properties – forged parts are generally stronger than machined parts;
- hardness – hardened parts have higher surface hardness;
- toughness – case-hardened parts with hard surface are still tough;
- form – some complicated forms (shapes) can only be achieved by casting;
- material – only a limited range of materials can be cast, or machined;
- surface and tolerance – the method of machining must be selected according to the specified tolerance and surface finish;
- dimension – the size of a part influences the choices of method and means of manufacture.

The design properties are also related to production properties and the resulting quality of manufacture. It is not sufficient if a technical system is well designed, and that this is verified by adequate design audits and technical and economic calculations. If it is to fulfill its functions, a technical system must also be well built.

The manufacturing properties of a technical system are merely the pre-conditions for rational and economic manufacture. Production itself is a separate problem, because in some senses this process is also a creative one. A good work team can, under favorable circumstances, manufacture a good product even with somewhat deficient design documentation. Conversely, a poor work team can manufacture a product that is merely acceptable, or even unacceptable, even with exceptionally good design documentation. The name of the company in some senses characterizes for the potential customer the expectation of quality of its products.

Designers must assess the influences that can reduce the quality of a technical system during production. These influences are the effects of the operators of the manufacturing process, as mentioned in appendix B. Designers must adjust the instructions on the design documents (drawings, parts lists, assembly instructions, adjustment rules, etc.) to these factors, and must consider whether manufacturing personnel are more or less qualified and experienced in the field. Naturally, contacts between designers and production organizations are important for the resulting quality of design and of the product.

7.2.7.10.8 Design Characteristics

The term "design characteristic" has appeared frequently in this book, in various contexts where its importance has been emphasized. This subsection serves to define that term again, particularly as an "internal" characteristic of technical systems that characterize and classify them for the expert [115]. Referring to the classification of design properties as presented at the beginning of this section, the design characteristics comprise the groups of special and general design properties. The third group, the elementary design properties, are the subject of the preceding sub-sections (10.1) to (10.7). The term design characteristics is used inconsistently and with various meanings in engineering practice and the literature.

A few examples should help to explain the term design characteristics as understood in this book. Let us analyze the following statement: "This is a reciprocating machine with four cylinders in line, it works on the four-stroke principle with direct fuel injection, has a certain compression ratio, and is water-cooled; it uses a wet cylinder liner, a confined gasket between cylinder block and head, full-skirted piston, rolling bearings; its first critical speed is above the normal operating maximum." All of these characteristics specify the machine for the expert in various ways. Some of the modes of action and operational properties of the TS can be inferred from such descriptions, but also some of the properties and particular features of the organ structure, the component structures, and others.

These examples also serve to explain another term used in connection with properties and design characteristics - their *state of embodiment.* Analyzing the statements about the motor, we are frequently induced to ask "what other possibilities exist with respect to reciprocating machines, four-stroke cycles, compression ratios, confined gaskets, rolling bearings, and so on?" We see that this example has named not the general design characteristic (number of cylinders), but a particular state of embodiment (four cylinders) as the description of a particular technical system.

From their nature, the design characteristics lie between the external (usage) properties that can be regarded as the goals, and the elementary design properties that serve as the means to achieve the goals. This discussion will be extended in section 7.3 to show that this position and the relationships to the other two groups can be clearly demonstrated, refer for instance to figure 7.16.

Only rarely can these relationships be expressed in mathematical equations, because they reflect a certain technical experience in addition to the modellable "hard" scientific content.

Figure 5.12 shows a systematic arrangement of some design characteristics, correlated to the four areas or domains [28] of transformations, functions, organs and components (compare figure 5.4). This list does not contain a complete enumeration of design characteristics for technical systems, but only those that generally occur as variety-forming characteristics in all types of technical systems.

Design characteristics can be related in a hierarchical way [31], usually given by a definition such as "the structure determines the elements and their sequence." Thus a design characteristic of higher order determines a sequence of design characteristics of lower order. For instance, a function structure contains design characteristics from the area of transformation systems (e.g. technology, output effects, and participation of a TS in their realization), and determines others (e.g. input, transformation functions, or groupings of functions).

7.2.7.11 Economic Properties (Ec).
The economic properties unite certain aspects of many of the properties described earlier into a few characteristic numbers. These represent the costs (expenditures) and the returns (income, benefits). The costs originate from the manufacturer, as acquisition costs (first costs) for design, development, manufacturing preparation, production, sales and distribution, results of complaints and reclamations, and as operational costs (running costs) to be carried by the user in performing the useful transformation processes. Both together are referred to as "whole-life costs". The returns of the working process arise as benefits from the transformation, i.e. the achieved total effect (see also figure 4.8).

To evaluate the economics of manufacture one must establish the self-costs of the technical system. The elements of this cost are shown in figure 7.10.

The self-costs form the most important economic property of a technical system from the viewpoint of the manufacturer. They consist of all the costs directly attributed to the manufacture of the technical system. These include not only the direct material and labor (personnel) costs, but also any determinable direct special costs such as some research, design and development costs, costs for work preparation in manufacture, and other dedicated expenditures, for instance for large-scale electricity consumption for electro-plating. If it is not possible to allocate these costs directly to the manufacture of a particular technical system, they should be included in the common costs (overheads).

Many interesting aspects of engineering design are shown by analyzing and evaluating those individual costs that are necessary to achieve the required properties for a whole technical system and its parts. These are shown not only by the absolute values of costs, but also by the mutual relationships among costs. For instance the proportion of material and labor costs for various branches of engineering can be very revealing. The range of variation in the proportion of total costs allocated to material costs is shown in figure 7.11.

a

Price	Self–costs	Produc-tion costs	Material direct costs (manufacturing material)
			Material overheads
			Labor direct costs (production wages)
			Production overheads
			Direct costs for production (e.g. licence fees)
		Design and development overheads	
		Administration and distribution overheads	
		Direct costs for distribution	
	Various supplementary loadings		

b

Price	Self–costs	Produc-tion costs	Manu-factur-ing costs	Machine hourly rate x machine piece time
				Labor wage costs
				Social overheads
				Other production overheads
				Production materials
				Material overheads
			Possible direct costs for production	
		Design and development overheads		
		Administration overheads		
		Distribution overheads		
		Direct costs for distribution		
	Various supplementary loadings			

Fig. 7.10 Cost Calculation Schemes

Self-costs are calculated from pre- and post-manufacture calculation formulae. Scheme a of figure 7.10 shows the cost-items that may be used for a surcharge calculation, scheme b contains the items for a cost center calculation; the latter is more accurate, but more difficult to handle (see also [18, 23]).

For the usual surcharge calculation, the direct labor (personnel) costs are used as the basis for calculating the overhead costs. A percentage surcharge to the labor costs is established on the basis of the end-of-year balance of the company, assuming that a predictive extrapolation to the current year (or month) is valid.

The self-costs serve as starting estimate for establishing the recommended price, which is probably the most important economic property of the technical system. The market price and the achieved sales price may be very different from the recommended price.

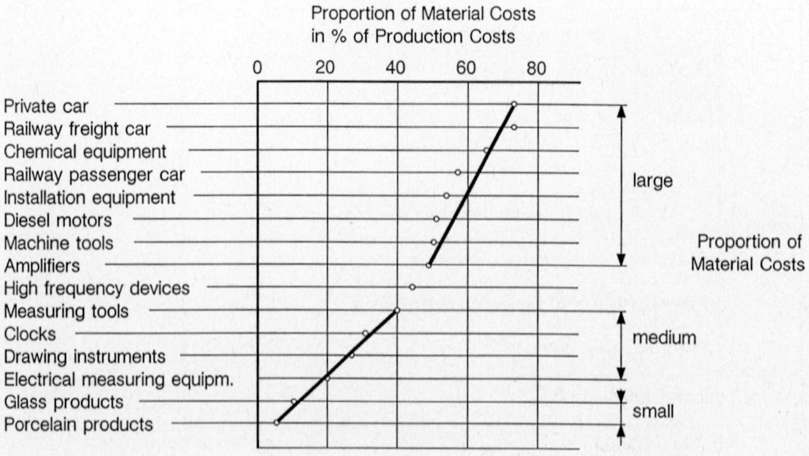

Fig. 7.11 Proportion of Material Costs in Production Costs of Some Products

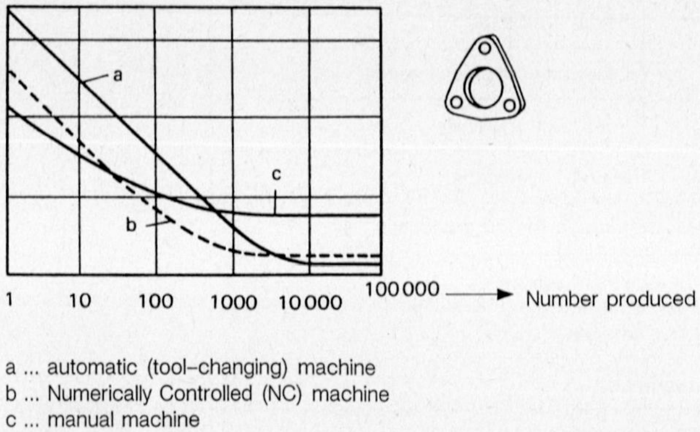

Fig. 7.12 Influence of Manufacturing Process on Production Costs - based on a simple machine component

The number of nominally identical technical systems to be manufactured influences both the labor costs and the manufacturing common costs (overheads), see figure 7.12. The manufacturing costs generally fall with rising quantities, and are influenced by the productivity of the operation and its personnel. For batch or series manufacture, the optimal batch size is important, see figure 7.13. This figure shows an idealized model of the relationship, based on linear assumptions (overheads completely independent of production quantity, and constant for a given time period; direct costs strictly proportional to quantity) [47]. The resulting optimum is shallow, a fairly large deviation from the optimal batch size (upward, or to a lesser extent downward) shows only a small penalty in increased costs. Other less tangible factors generally indicate that a smaller batch size is preferred.

The manufacturing costs depend to some extent on the number of parts, subassemblies and groups that are *re-used*, i.e. that do not need to be designed for the

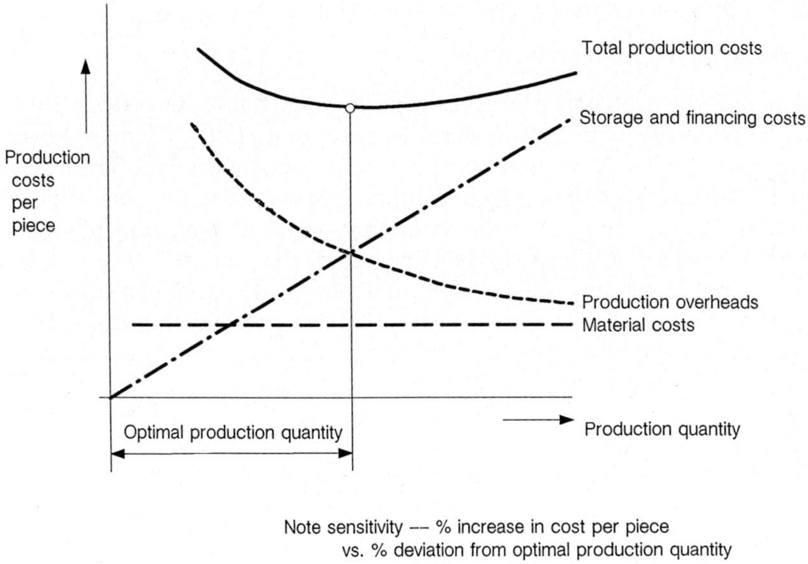

Fig. 7.13 Optimal Production Quantity per Batch

new technical system. Categories for classifying parts and groups from this viewpoint are illustrated in figure 6.7. The magnitude of the pre-calculated manufacturing costs serve as one criterion for the manufacturer of a technical system whether it is more economical to manufacture in-house, to purchase them from another supplier, or to contract for their manufacture.

The operational (running) costs that arise from the technical system during its working process are important for the user. Their most important component is usually amortizing the initial cost of purchase over the useful life of the technical system. The operational costs are calculated from the purchase price, costs of transport to the operating location, and all other costs connected with making the technical system operational, e.g. energy provision, preparation of the operational base including foundations, commissioning and acceptance tests, etc.

The other components of operational costs are contributed by the operands and operators of the working process, see figure 4.1.

The operational costs are contrasted to the economic and societal benefits during the same time period. Benefits are expressed either in technical or in monetary units. The ratio of benefit to operational costs is termed *effectiveness*, and is one of the most important economic indicators (see also section 4.7).

EV = Benefits of Process / Expenditure for Transformation

or:

EV = Working Effect of Process / Running Costs of Process

in the same time period.

Another important economic indicator is *profitability*, which relates mainly to financing of an operation. It can be defined as:

PV = Profit in a given time period / Capital employed

also referred to as "return on investment".

Calculation of the economic properties consumes much time and effort. There are many imponderable factors involved in such predictions, such as future changes in rates of interest, taxation policies, subsidies and incentives. The resulting values are usually referred to a particular time by calculating the time-value of money, and including estimates of loss of interest on investments, including the inventory that is needed to support production and distribution. It is little wonder that designers are usually not familiar with such calculations, and frequently tend to neglect the economic properties during the important origination stages of a technical system (design and development), thus at times jeopardizing evaluations, choices of optimal variants, and marketability.

7.2.7.12 Liquidation Properties (Liq).

At the end of life of a technical system (see also section 7.2.7.3), the TS is normally taken out of active operation, and liquidated. Liquidation can take place in various ways, and provide certain requirements and constraints to designers, manufacturers and users. Liquidation can result in waste or scrap.

Waste is defined as a permanent disposal in which the whole or parts of the TS are not used again. Usually, some storage location must be found such that the waste does not constitute a hazard. Landfill sites may be provided for this purpose. Various hazards can exist that demand some action from the liquidator, for instance (a) preventing entrapment by removing covers or doors from containers (freezer chests, refrigerators), or welding them closed, (b) removing fluids that can be polluting or poisonous (coolants, acids, oils), which may involve cleaning operations before disposal, (c) encasement for long-term hazardous substances (nuclear wastes, chemicals), etc. A prime concern must be for preserving the environment, which in this case can easily include the earth's oceans or atmosphere.

Scrap usually implies that some dis-assembly takes place, and some parts of the TS are re-used. Such re-use may involve (a) salvage of individual components or groups to be repaired, re-furbished or renovated and re-used as replacement or spare parts, (b) recovery of constituent materials, and re-cycling to upgrade them into re-usable raw materials (or intermediate material forms), thus adding to the available resources, or (c) archival applications, as exhibition pieces in museums or historic landmarks.

In some instances, certain (parts of) technical systems can be used for *energy recovery*, either by natural processes such as fermentation, or by combustion. The energy can be made available by collecting combustible gases or liquids, or by waste-heat recovery techniques from the combustion process.

Designers must also consider what will happen to the TS during this phase of its life, they may be held liable in law for any damage that may result from their products.

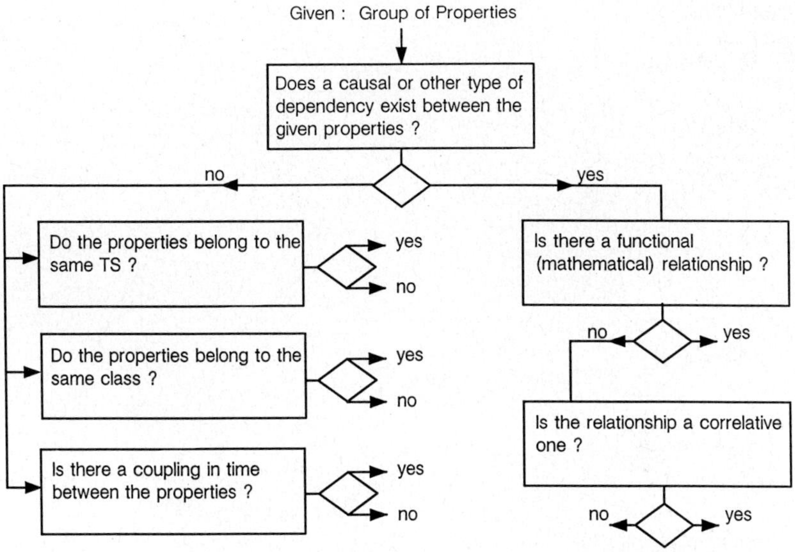

Fig. 7.14 Determining Dependencies between Properties or Groups of Properties

7.3 Relationships Between the Properties

A general definition of relationships is given in appendix B. A systematic overview of relationships among properties can be guided by the flow chart in figure 7.14, where the outputs labelled "yes" represent classes of properties.

Properties and their general relationships are treated in various fields of natural science and engineering, and relationships expressed either in equations (or inequalities) or words. Mathematics, including mathematical logic and statistics, is an important discipline for studying relationships. It offers possibilities of expressing relationships in symbolic form, and using formal operations to evaluate them. Very complicated relationships can be expressed, for instance, in matrices.

The general relationships are usually not sufficient to achieve the required effects from a concrete technical system. They must be adapted to the existing conditions to take account of the complicated phenomena that appear in practice. For instance, heat transfer can be modelled by mathematical expressions for conduction and radiation, and to a lesser extent for convection. Which relationships are valid and sufficiently accurate for a workpiece that is placed in a heat-treatment furnace? To solve such cases, designers must have experience in the relevant fields that permit them to reliably estimate the concrete conditions of the special situation. Even the most recent developments of digital computation by approximation methods, finite elements, finite differences, boundary elements, and a host of others, can only give an indication within the limits of accuracy and resource expenditure of the available computing machinery. Much of the applied research can also only yield empirical equations of limited validity (e.g. [85]). Relevant experience can be expressed in correction coefficients which can then be applied to the general formulae.

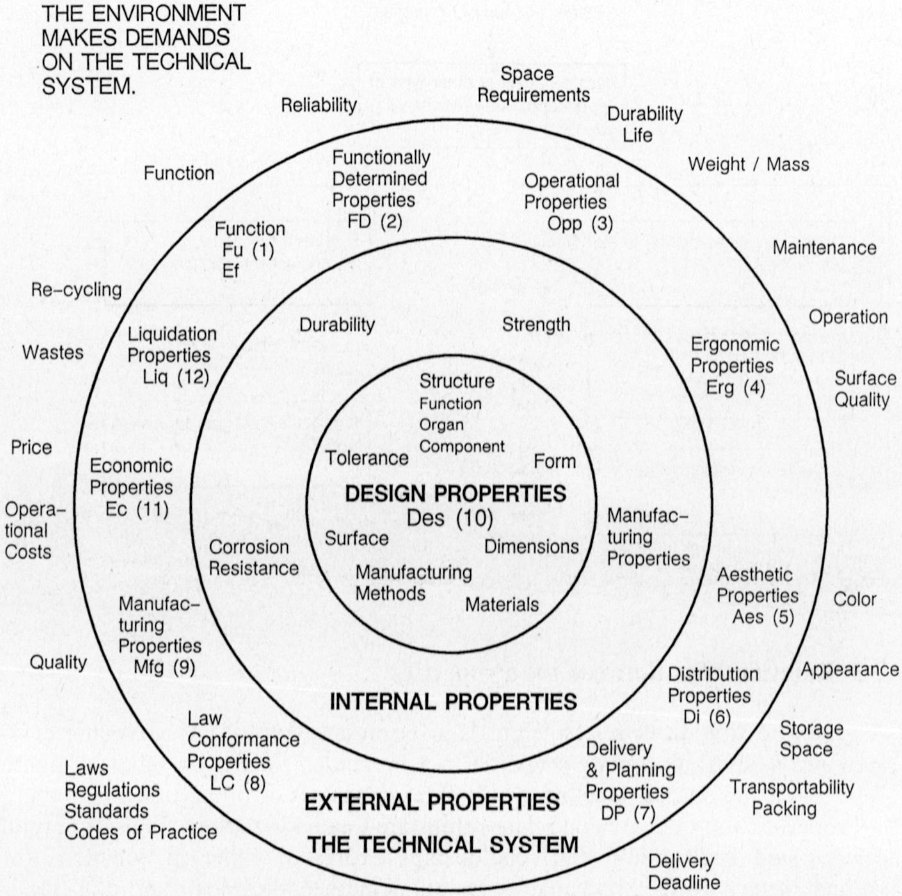

(For explanations, see also figure 7.1)

Fig. 7.15 Relationships between Classes of Properties

It is also important to comprehend the relationships between the categories of properties. For a finished product, figure 7.15 shows that the elementary design and manufacturing properties located in the central circle, and the adjoining ring of internal properties, "support" the external properties, which in turn determine the economic properties of the system. The "external pressure" of the requirements is resisted by an internal "excess pressure" created by the quality of individual properties. How these categories depend on one another is illustrated in figure 7.16. A different view of these relationships is given in figure 7.17 [210].

The examples presented above reveal that these relationships are very complicated, and this complication is increased even more by the relationships between technical systems of lower degrees of complexity.

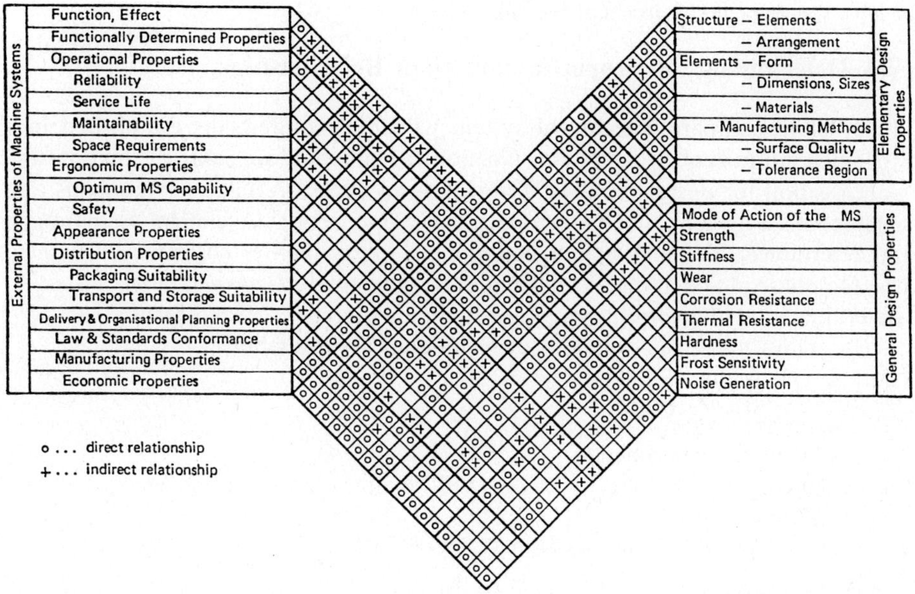

Fig. 7.16 Relationships between Classes of Properties and the Design Properties (Reproduced from V. Hubka, Principles of Engineering Design, London: Butterworths, 1982 [119] by kind permission of the publishers)

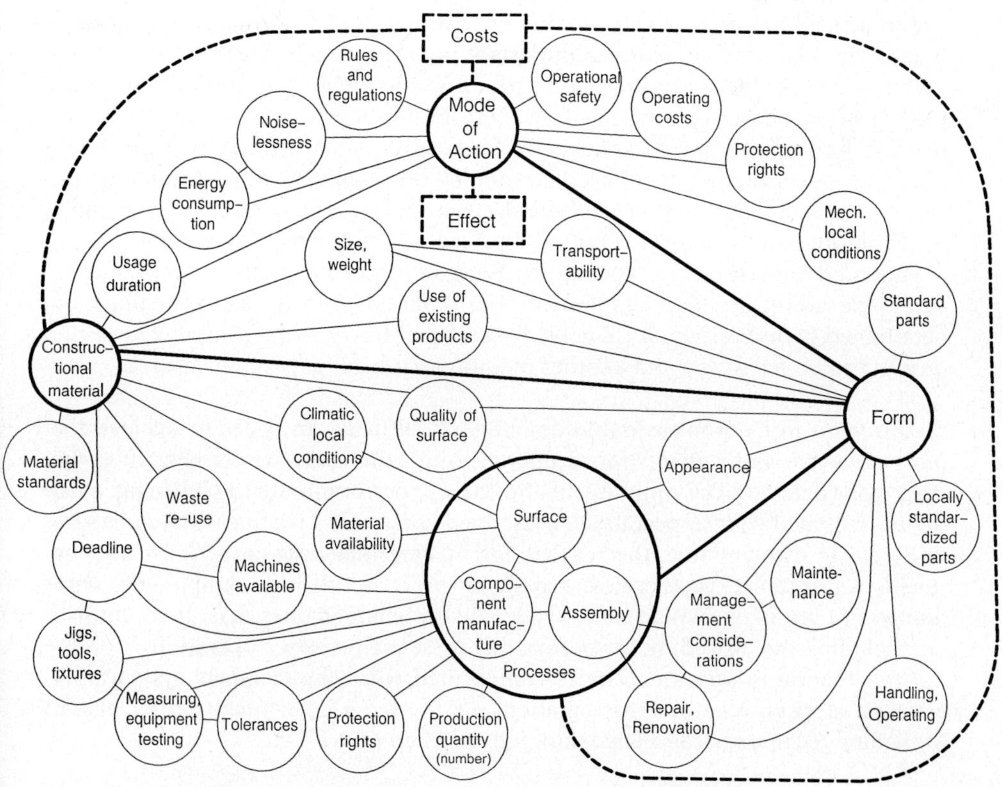

Fig. 7.17 Relationships between Various Properties of Machine Systems [209]

7.4 Determining the Properties and Their Relationships

In order to compare a technical system with the requirements established in a specification (a requirements specification for purchase or contract, or a design specification for designing, see figure 5.4, level O), or to make comparisons and evaluations between technical systems, the properties of a technical system must be determined. The methods of determining the properties change according to the life phase in which this determination is to be made, i. e. whether the system to be evaluated is in a conceptual phase, or in a materially realized form.

The following basic operations or techniques are available:

(1) measuring;
(2) estimating;
(3) modelling by experiments;
(4) calculating or reading off (from graphs, diagrams, nomograms, etc.);
(5) comparing;
(6) determining optimal measures (values).

(1) Once a technical system is realized, the quantifiable properties may be determined by *measuring*, and those properties that are not quantifiable may be estimated or assessed by experts. Even though details of these activities are the subject of chapter 8, some of the problems are highlighted here. Measuring is an expert activity that has its own working methods. As with all other activities, preparation for measurement plays an important role, both with respect to the working process of taking measurements, and with respect to the documentation and tables for recording and evaluating the data. Test establishments have generally developed various pre-printed documents for frequently recurring measurements. Standardizing organizations are concerned (among other duties) with formulating test procedures that permit consistent measurement of relevant variables, for example the Schlesinger procedures for testing the accuracy of machine tools [190].

Particular provisions are needed for long-time measurements, for instance to determine useful life of a component. The required property can sometimes be determined by indirect measurements based on a known relationship between the quantity to be measured and another quantity that is more easily measured.

(2) *Estimating* of non-measurable quantities is difficult. In order to achieve the most objective assessments, the properties and criteria for assessment must be accurately defined. The opinions of the experts performing the assessment must be substantiated with respect to the prescribed criteria. If a technical system is said to be easily maintainable, the reasons for making that assessment must also be stated, for instance that critical parts are easily accessible, components with limited life are easily replaceable, service by the manufacturer is assured, and so on, including the procedure for performing these maintenance operations.

The situation is very much more complicated when the technical system only exists as a rough idea or representation, e. g. during a conceptual design phase. Modelling techniques can be meaningfully applied in these cases.

(3) *Modelling techniques* attempt to represent or simulate the technical system. An

"investigation" of an abstract technical system is inordinately more difficult than measurement on a physical reality. Where the designer or expert can support the ideas by experience in similar situations, the conclusions are relatively easy to reach. If the situation is novel, and originates from different or new conditions, modelling techniques can be used for objective assessments.

A model is a representation of the real (realized or proposed) technical system, the process, or the idea (or hypothesis) by suitable means. The relationships between a model and the original are generally the laws of similarity (see appendix B). Modelling techniques have been expanded in recent times with respect to both widened usage and extent of the investigations by application of digital and analog computers.

Clarity must be achieved about the viewpoint from which similarity between the model and the original is of interest, i.e. which properties are to be expressed in the model, and what purpose is to be served. A prototype of a technical system permits determination of most of the properties relevant to the final system. In contrast, a model only permits determination of certain properties, such as behavior, structure or form, which lend their name to the appropriate model.

A model always has a definite purpose. It may not be just for determining the measures of properties, but may serve also as means for checking and verifying, communicating, or instructing. Aspects of models can be summarized by a model of models [171] shown in figure 7.18, which demonstrates four dimensions of models qualities.

CONTEXT	FUNCTION	MEDIUM	MODE
Material Concrete Conceptual Abstract	Describing Predicting Exploring Planning & Designing Prescribing Heuristic Normative	Verbal Mathematical Symbolic Imagal Graphical	Iconic Similitic (used as signs) Metaphoric (used as symbols)

Fig. 7.18 Model of Models (adapted from [170])

(a) **context** ranges in a wide spectrum of variety from abstract to concrete, from material to conceptual, from general to specific;
(b) **function** and purpose can be one or a combination of the following:
- describing, to explain some aspect of the model and the reality,
- predicting, to foresee some aspect of behavior, and to quantify it,
- exploring, to investigate behavior under possible changes of circumstance,
- planning and designing, to propose new or novel applications or devices,
- prescribing, as a normative or heuristic instruction;

(c) **medium** can be one or a combination of:
- verbal, using words and word structures,
- mathematical/symbolic, using symbols to represent and manipulate the model,
- imagal/graphical, using two- and three-dimensional visual images or graphical representations;

(d) **mode** of usage can be:

- iconic, representing the reality in a two- or three-dimensional representation that is usually immediately recognizable, frequently in an enlarged or reduced scale. Models in this mode of usage include drawings, space models of machines or workshops, photographs, but also certain forms of mathematical equations and realistic verbal descriptions, where the similarity between reality and model are very noticeable,
- similitic, or analog, in which only a limited number of properties of the model are similar to the real system. Static and dynamic properties of a reality can be imitated or simulated by this use of models. This usage group includes graphs and diagrams, but also models that rely on laws of similarity, e.g. for fluid flow, electrical or magnetic fields, thermal conduction, digital computer simulation programs, and others,
- metaphoric, where a mathematical, verbal or graphical symbol is used to represent a context. For instance, this book frequently refers to a technical system by using the abstract, descriptive, verbal, metaphor "TS".

The relationship, especially in context and function, between a mental model and the modelled reality is shown in figure 7.19. Mental models are formed by abstractive documentation, by perceiving and abstracting from a physical reality, and their properties are elaborated by generalizing and theorizing. When a physical reality is being operated, the mental model is applied to the reality in a 1:1 situa-

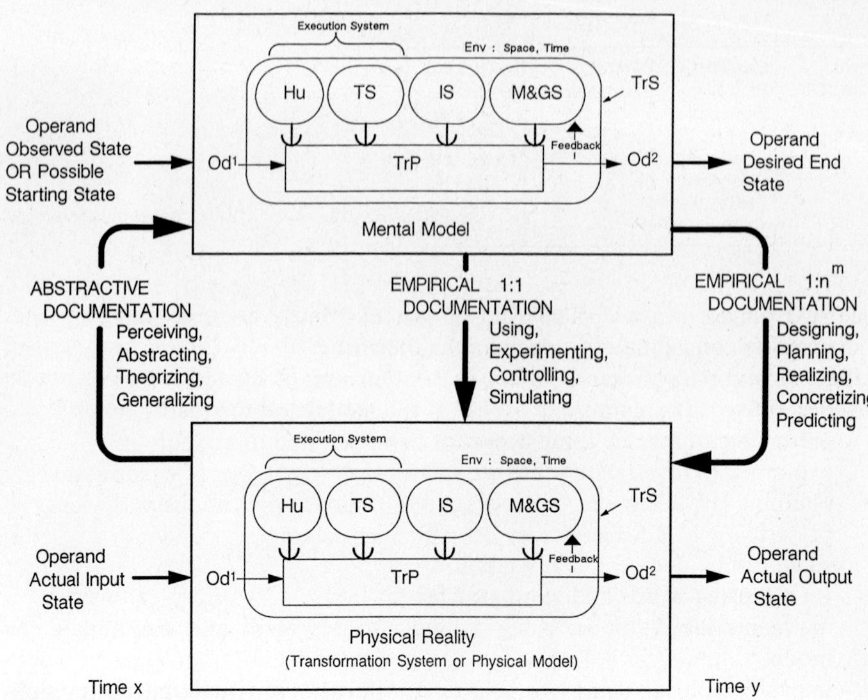

Fig. 7.19 Formation and Use of Models (adapted from [64])

tion of empirical documentation. The purpose of operation may include using, experimenting, controlling, simulating, etc. If a physical reality does not exist at the time, the empirical documentation must be developed by planning, designing, realizing, concretizing, predicting, etc. Each of these operations usually takes place in a number of stages, and some alternatives are available at each stage. The resulting combinations may reach large numbers, of which only a smaller number may be viable and usable. The subject of designing will feature in chapter 12.

The relationships of similarity between two systems are investigated by the *laws of similarity*, whereby one can distinguish between physical and mathematical similarity. The laws of similarity are based on the concepts of dimensional analysis and the Buckingham Π theory, through which it is possible to express almost all physical processes by means of dimensionless (or unity-dimensional) groups of variables (compare [42, 60, 61, 155, 174, 189].

The most usual purposes of *experiments* on models are to confirm or reject a certain hypothesis, or to determine the basic data for quantifying a relationship. Experiments should be performed according to prepared plans which specify the changes in variables and experimental conditions, and prescribe the processing of results according to deterministic or statistical rules (refer to [41, 49, 97, 163, 189]).

(4) If diagrams such as graphs or nomograms exist, the desired measures of properties can be determined by *reading* them off. Those properties represented by equations can be determined by *calculation*, which includes usage of analog and digital computers, and exact and approximate methods of calculation.

(5) A reference sample can also serve as a model. For instance, samples of surface quality for different machining processes are available. The measures of the relevant properties can then be determined by *comparing* the object to the samples. Colors are usually measured in this way, and color comparisons can be used to measure other properties, e.g. acidity or alkalinity (pH) using a chemical indicator.

(6) *Determining optimal measures of properties* under certain given conditions occurs frequently as a problem in engineering. Examples of such measures may be the optimal speed of a vehicle in traffic, optimal power of a motor, optimal form for an airfoil, or optimal number of bolts for a flange connection. In order to determine an optimal value, the available knowledge about relationships between properties is used. A usual situation shows that one group of properties is positively influenced, and another is subject to a negative influence. The properties to be optimized may change their measures:

- discontinuously, by integer values such as number of struts, or gear steps, or by stepwise changes in relationships;
- continuously, such as insulation strength, or vehicle speed.

Figure 7.13 serves as an example in which the optimal order quantity for reordering a batch-produced item is to be found in order to minimize costs. It is known that the expenditures for wages and overheads generally fall with rising quantities, but the material costs remain approximately constant, and therefore directly proportional to batch size. Additional costs will occur for inventory storage, transport and financing because the parts will only be used during a period of time. By adding all the costs one obtains the batch size for which the manufacturing costs are a minimum. The sensitivity of this minimum to deviations from the optimum should also be investigated.

The task of determining an optimal value can be performed by graphical or numerical calculations. Analytical methods are broadly used, especially on computers. They allow calculations that were previously uneconomic because of the large time investment needed for non-computer methods.

Calculation processes for determining an optimal measure (mathematical optimization) are generally well known, and developments are reported in the literature, with applications in many fields [50, 80, 129, 145, 160, 177, 196]. Nevertheless, a brief review of the general procedures is given here, under the usual assumption that the criterion for an optimum is represented by least cost of acquisition (or self-cost) or maximum effectiveness (see section 4.7 and 7.2).

(a) For parameters that vary *discontinuously*, especially for integer variation, the optimal case is determined by calculating the self-costs or effectiveness for every technically feasible case. The best of these is the one that shows lowest self-cost or highest effectiveness. Optimizing can take place by calculating one case after another, and collecting the results in a table or diagram that can reveal trends. In more complicated cases, the computational methods of linear, non-linear, dynamic, or geometric programming, and other varieties of mathematical optimization can be used. Reference should be made to the literature for details of these methods and their applications and limitations.

(b) Determination of an optimal value for a *continuously* variable parameter is performed with the help of an analytical method. Examples include calculating the optimal speed of a transportation medium, optimal voltage for electrical energy transmission, or optimal diameter of a pipeline.

Initially, all properties that are recognized as having a significant influence are expressed as a function **x** of the parameters to be optimized. This is transformed into an expression of effectiveness **EV** as a function of the combined parameter **x**.

$$\mathrm{EV} = \mathrm{F}(\mathrm{x})$$

If this function is continuous, and has continuous derivatives, the parameter **x** has an optimum $\mathbf{x_o}$ when the function **F(x)** shows an extreme value. The value of $\mathbf{x_o}$ can then be calculated from the relationship:

$$\mathrm{F'(x)} = \mathrm{dEV} / \mathrm{dx} = 0$$

The conditions for **EV** to distinguish between a maximum and a minimum is that:

$$\mathrm{F''(x_o)} = \mathrm{d^2EV} / \mathrm{dx^2} > 0 \rightarrow \text{maximum}$$

$$\mathrm{F''(x_o)} = \mathrm{d^2EV} / \mathrm{dx^2} < 0 \rightarrow \text{minimum}$$

In most cases the function can be readily differentiated. The resulting expression of the first derivative that is set equal to zero is frequently a higher-order equation, such that solving for **x** is not generally possible. Such cases may be solved by an approximate numerical solution, usually with the aid of a digital computer.

(bb) A similar procedure is used for determining the optimal values for each of *two continuous variables*, especially if the parameters cannot be independently optimized, for instance in the case of the optimal cross-section and life-span for a pipeline.

All relevant properties are expressed in a function of the two variables **x** and **y**, e.g the costs as function of the cross-section and the life of a pipeline. This yields the effectiveness as a function of the parameters to be optimized:

$$EV = F(x,y)$$

The first partial derivatives of the function **F(x,y)** with respect to each parameter **x** and **y** are:

$$F'x(x,y) = \delta EV \,/\, \delta x \mid y$$

$$F'y(x,y) = \delta\ EV \,/\, \delta y \mid x$$

If both partial derivatives are set equal to zero, a system of two simultaneous equations for both optimal values is obtained:

$$F'x(x_o,y_o) = 0$$
$$F'y(x_o,y_o) = 0$$

From these two equations, the optimal values $\mathbf{x_o}$ and $\mathbf{y_o}$ may be determined by numerical or graphical calculation. Checking the second derivative can confirm that the values are for a maximum effectiveness.

All of the operations to determine measures for properties outlined in this section are elementary operations of the design process, as will be briefly discussed in chapter 12, see also [119].

7.5 Specification of Properties of Technical Systems

When speaking of a “list of requirements” or a “problem assignment” the intention is always a specification of the desired properties of a technical system. Such a specification can take a number of forms, depending on the purpose of the issuing persons or users, compare figure 5.4, level O. A *requirements specification* is usually issued by an official body (e.g. a government office) to state what requirements must be fulfilled, and the circumstances of acceptability of a product. A similar but less formal specification is issued by potential customers as a *request for tender.* On awarding a contract to design and construct a technical system, a formal *design contract specification* is usually approved and signed. A designer will try to understand a design problem that has been assigned to him by writing a *design specification.* These specifications relate to technical systems that do not exist at that time. Other specifications relate to existing systems, for instance manufacturing specifications, test specifications, operation specifications (e.g. user manuals), and so on.

An incomplete specification accepted in the early stages of request or design is frequently the cause for errors and faults in the finished product. It is almost irrelevant whether such omissions are faults of negligence, or whether these items were regarded as self-evident and obvious. The importance of a correct and complete design specification is underscored by the saying: A correct problem statement is already half a solution.

The contents and form of a specification are different from case to case, and are influenced by a number of factors:

(1) complexity of the functions - complete plant vs. machine component (see sections 6.1 and 6.3);
(2) design difficulty - new design and development vs. adaptive design (see section 6.5);
(3) requirements for additional properties - safety, life, appearance;
(4) problem initiator or sponsor.

The first three of these factors have been discussed in previous sections of this book, particularly in chapter 6. In order to relate these factors, consider figure 7.20. Products generally progress with time towards higher complexity and matu-

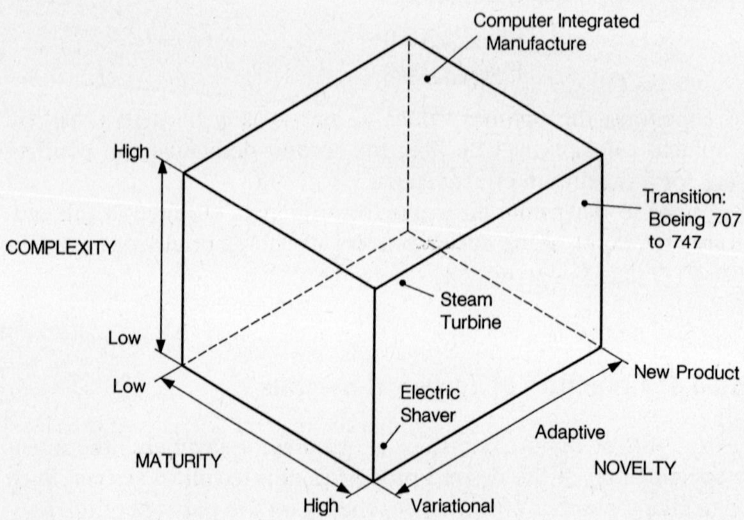

Fig. 7.20 Relationship of Design Factors and Products

rity but lower novelty. Risks of failure of the designed device usually increase directly with novelty and complexity, and decrease with maturity. Some added complexity may reduce the overall risk of failure, for instance when safety or warning devices are added to existing products, but they also add new elements of risk. Maturity manifests itself among other features in established codes of practice, laws and standards, quality and quantity of empirical and scientific know-how and knowledge, and small evolutionary changes in the product. Novelty manifests itself in unconventional ideas, features and conceptual combinations that "are not obvious from the state of the art" (the definition accepted for patentability), and that are incorporated as innovations into a usable and marketable product. A novel, immature or complex product requires much more thought and prior work remote from the drawing board, in the conceptual phases of designing, before entering the lay-out and detail design phases. It also requires much more and better management, than a simpler, maturer and better established product.

Factor (4) mentioned above, the problem initiator or sponsor, is related to *satisfaction of needs* (compare also chapter 2). Two possibilities exist, see figure 7.21:

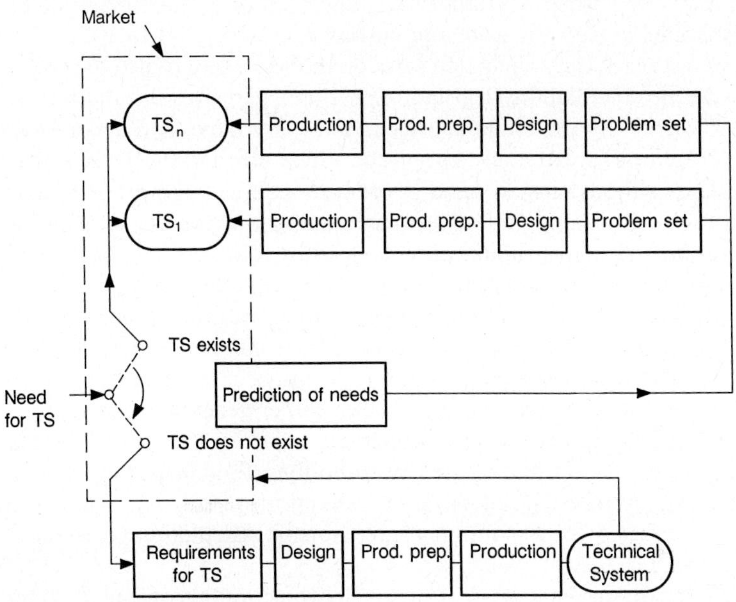

Fig. 7.21 Types of Need Satisfaction

(a) Purchase of a finished product. Consumers who wish to solve their problem by means of a technical system search the market to find whether the desired product with the required external properties and parameters exists. Normally a choice between a number of products is available. This is the most usual case.

(b) Ordering a special product. If the technical system with suitable external properties and parameters is not available on the market, customers must wait until the long process from problem statement and assignment through manufacture and installation has been accomplished. Customers then obtain a product that meets their particular requirements. The high price makes such products attractive for the manufacturer to improve turnover and returns on investment.

With respect to designing and manufacturing, the persons who assign the problem, the sponsors, are different in these two cases. In the first case, a potentially anonymous customer must be represented by "someone" who anticipates the requirements, needs and demands, and formulates the design task in a specification. This is usually done by the sales and marketing departments of a manufacturing organization acting as the sponsor. In the second case, for direct ordering of a new product, the customers or their representatives are the sponsors, and assign the problem.

In direct ordering, customers can validate their requirements directly to the manufacturer, and must thus accept the risk of formulating the problem assignment. The manufacturing organization guarantees that the technical system exhibits the required properties, and thus carries "only" the technical risks.

Where the formulation of the task occurs within the manufacturing company, from a sales or marketing office, the danger may arise that the requirements as formulated do not coincide with the potential customer's vision, and that therefore the product does not reach the anticipated sales quantities. There is also the risk of court action under product liability legislation, if a user perceives an undisclosed hazard. The risk carried by the manufacturer is thus very large, and accordingly the definition of the task is extremely important. Great effort must be expended and all appropriate methods must be used in order to assure a high probability of success. This requires close cooperation between a number of experts and specialists, among whom the designers should play the leading role.

Because a requirements specification can never be complete, designers must establish requirements for additional properties for the technical system which is intended to solve a design task. The customers (or their company-internal representatives) usually state their requirements within the range of the external properties of a technical system. Only certain types of customer generally state requirements that concern internal or design properties, such as the armed forces, national postal service, rail services, regulatory authorities (food inspectorate), and major insurance organizations (Underwriters Laboratories, Lloyds). These can include materials, surface qualities, minimal safety factors, calculation procedures, and so on.

All the above refers to the final product that is delivered to the consumer. A different situation exists with respect to partial systems and elements, where designers formulate the requirements for themselves.

The questions listed in the various categories of properties, figure 7.1, can serve to ensure completeness of the design specification. Starting from those categories, a more specific list of properties can be generated for a more concrete group of products, especially as the number of properties is relatively more limited.

A particular case of requirements comprises the conditions that may be stated to limit certain properties, *constraints.* Some of these constraining requirements are imposed on the designer by nature such that certain solutions are not issued. We do not usually regard these as constraints, we have accustomed ourselves to the laws of nature and acknowledge that they cannot be broken, for instance that a *perpetuum mobile* cannot be made. A second type of constraint comprises the decisions of society and authorities, including laws, regulations, codes of practice, codes of ethics, etc.

All design requirements must be classified (by designers or sponsors) according to their *importance* (priorities) and the opinions of the customers (or their representatives), typically into three groups :

(1) Requirements that must be fulfilled under all circumstances, such as those respecting function, and usually also price. These can be divided into *fixed requirements* (demands) that must be fulfilled within a tolerance, *maximum requirements* that shall not be exceeded, and *minimum requirements* that must be exceeded;

(2) Requirements that may in exceptional cases be relaxed, or not completely fulfilled, or reviewed with the sponsors;
(3) Desired requirements (wishes) that should only be fulfilled if the circumstances are favorable.

Designers must always be aware to which of these classes a particular requirement belongs, and they should also have this classification confirmed by the sponsor, preferably in writing.

A further group consists of the *permanently implicit* requirements that are fulfilled by the designer without being specifically stated. It is part of the professional pride and ethics of an engineering designer to come to terms with these requirements. They include:

- The highest attainable measures of operational properties, and among them in particular:
  - minimal space required,
  - minimal weight,
  - minimal energy or fuel consumption,
  - accessibility and replaceability of all elements with limited expected life,
  - optimal reliability.
- The best possible ergonomic properties:
  - simple and easy operation,
  - protection from all secondary outputs such as noise, heat, vibration, exhaust gases, dust, etc.,
  - minimal danger to the environment.
- The maximal utilization of all resources of the manufacturing organization:
  - use of existing materials and documentation (components, component families, standard parts),
  - effective use of all means of production, including machines, jigs and tools, measuring equipment.
- The most favorable economic parameters:
  - minimal manufacturing costs,
  - minimal operating costs,
  - minimal reclamation costs (or minimal throw-away value).

In addition, a number of guidelines and recommendations can be derived from the above implicit requirements. These can be termed *working principles* for the designer (an extended collection appears as table 4 in [119]), for instance:

- simplest possible design solution,
- as far as possible equal life for all components, or easiest exchange of parts that are subject to greatest wear and tear,
- sensible utilization of strength properties of materials, especially at points where forces are applied,
- use of natural resources of the country or region, considering the current situation of the population,
- respect for well established and sensible traditions.

Many of these working principles overlap in their scope, and one must try to reach an optimal compromise among opposing tendencies. This problem will be re-considered later in this book. A humorous discourse on this topic is "The Deacon's Masterpiece, or the Wonderful 'One-hoss Shay', A Logical Story", by Oliver Wendell Holmes (1809–1894), from his book "The Autocrat of the Breakfast Table" (New York: Dutton 1906 and Heritage, 1955), and frequently reprinted in various anthologies.

7.6 Mental Processes of Realizing the Properties of Technical Systems

A remaining open question concerns the people who realize the properties of the technical system by their mental processes, engineering designers, and on what circumstances the values of the realized technical system depends. It is obvious that even the best lay-out on paper will turn out to be useless if the technical system is not realized (manufactured and produced) satisfactorily, and that without good organization for realizing it, the product will be delayed and the costs extremely high. But without the expertise, work and imagination of engineering designers it would be unthinkable to satisfy today's high expectations.

Designers are the creators of technical systems and of most of their properties. In contrast, their influence on the manufacturing, delivery and planning properties is less, but they create the prerequisites for realizing these properties. They take part in establishing the appearance and ergonomic properties, if necessary in cooperation with experts in aesthetics and ergonomics (e.g. product designers in North America, or industrial designers in England). They also have a large influence on the economic properties of the technical system.

The "values" of a technical system are progressively created during the design process. Some of the factors that influence the results of this process are discussed in chapter 12, compare also [119]. Only the basic factors that determine the value of a product are mentioned here. These are the expertise and technical knowledge of designers, the time taken for the design process, the number of persons involved in designing, and the number of improvements.

A qualitative graphical impression of the relationships between product value, expressed as the degree of fulfilling the property requirements, and the expertise and designers' knowledge is shown in figure 7.22 [210]. Unfortunately, the terms used to label the scales are not accurately defined, and therefore cannot be quantified.

Figure 7.23 shows qualitatively how the value of a technical system, expressed as the design maturity of the product, depends on the time expended in the design process, and the number of active designers, the personnel capacity for designing. This design capacity can for instance be expressed as a quantity of money per unit time. The diagram shows that after a time $\mathbf{t^x_0}$ the quality of the resulting product increases relatively more slowly, but that by increasing the capacity the time does not reduce in proportion. These statements are confirmed by observations of practice. The relationship shown in the diagram is, of course, only valid if the design team has a sufficient amount of expertise and knowledge available. A team with other qualities would show a different trend of progress. For a weaker team, for instance, the distance of the working curve from the asymptote of the ideal solution would be larger, and the slope of the curve shallower.

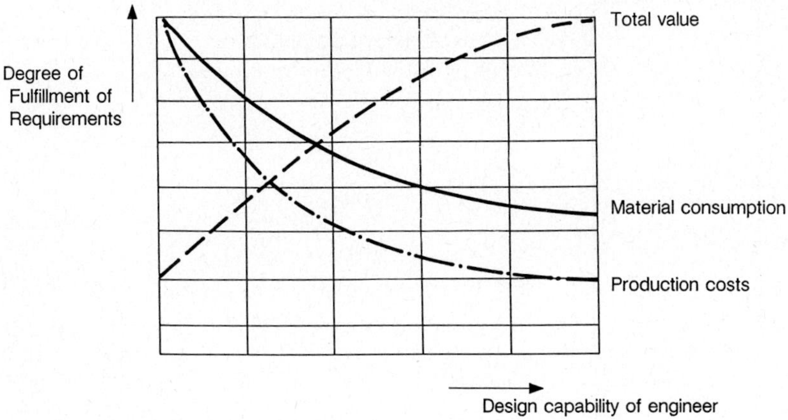

Fig. 7.22 Influence of Design Capability on Fulfillment of Requirements [209]

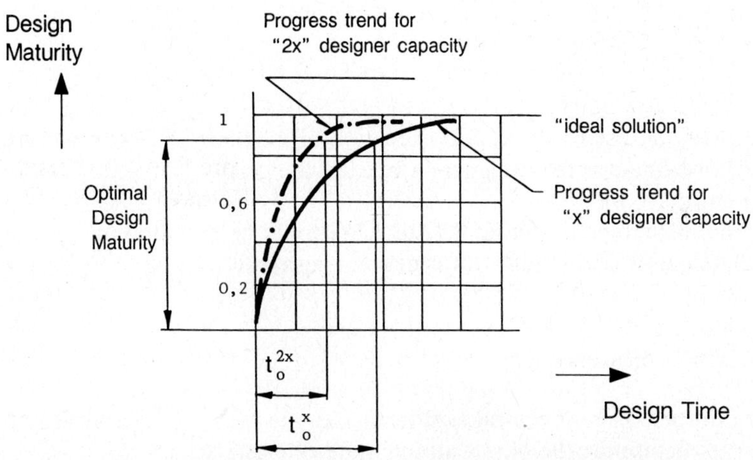

Fig. 7.23 Design Maturity Depending on Design Time and Capacity

At each re-design or review of a technical system (e. g. for further development), new measures of properties should be achieved. Such progress can be indicated on a "relative strength diagram", as discussed in chapter 8 (figure 8.3). Possibilities for improvement are always limited, either by laws of nature, or by the state of the art and scientific knowledge (to be discussed in chapter 11, especially figures 11.4 and 11.6).

PROPOSITION 7.9

The basic factors on which the value of a realized technical system depends are: the abilities of the design team (see figure 7.22), design time (see figure 7.23), and the number of improvements.

It is interesting to discover how much influence designers have on the manufacturing costs of a technical system. The common conviction is that these costs depend

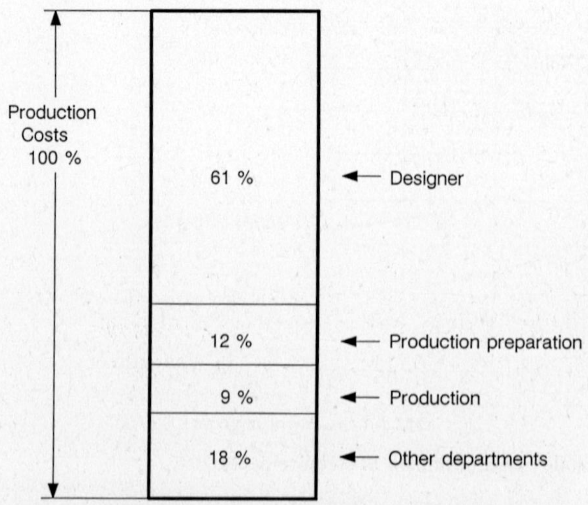

Fig. 7.24 Responsibility for Production Costs in One-off Manufacture

mainly on the planning, preparations and execution of manufacture. An accurate determination shows that it is the designers who exert the greatest influence on the manufacturing costs, see figure 7.24.

The information contained in figures 7.22 to 7.24 should be regarded merely for guidance. Much depends on how this information is interpreted.

7.7 TS Properties - Statements

The category of properties is of central significance to the theory of technical system, the properties determine the object and its state of existence.

Statement 7.1 Technical systems must possess their purpose function, the capability of action, and the capability of exerting suitable effects as their main output. They must also possess certain well-defined measures or states of embodiment of a finite set of properties which make technical systems suitable for use. Some of these properties provide direct support for the capability of action, others form additional aspects of the product.

Statement 7.2 A full listing and survey of the properties may best be obtained by considering the classes of properties and aspects (see figures 7.2 and 7.16)

Statement 7.3 Every technical system possesses all types of properties, whether these are consciously planned according to the requirements, or not.

Statement 7.4 The starting point of technical systems are the requirements that determine the properties of the future TS.

Statement 7.5 Properties may be categorized according to various viewpoints. Important classes are the external (usage) properties, and the internal (design) properties.

Statement 7.6 External properties depend on internal design characteristics and properties (see figure 7.16).

8 Evaluation of Technical Systems

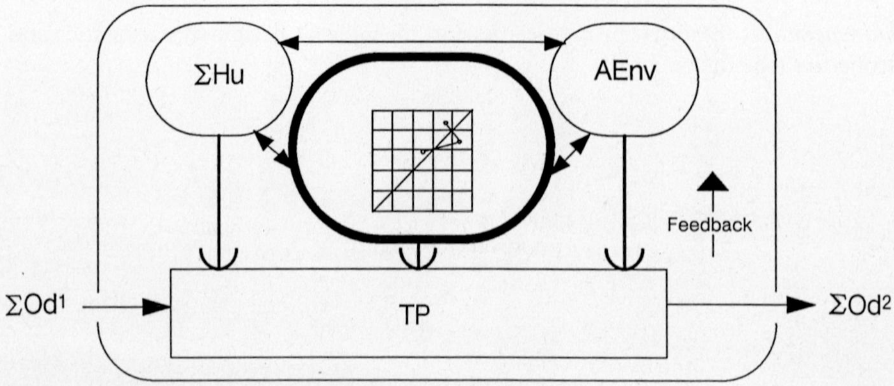

At the beginning of chapter 7, the question "how is the car?" was asked. This question implied "what qualities does it have, what properties and their measures, what performance?" In order to answer this question, certain properties must be selected and evaluated, and the resulting (possibly large) number of measures must be combined into a commonly understood characteristic value. Combining measures is not easy, especially if we are dealing with different properties such as speed, power, equipment in the passenger compartment, appearance, surface hardness and others. Only if such a meaningful combination, a *synthesizing characteristic*, can be achieved can we objectively assert that one car is better than another, because it has a higher characteristic value. Some of the problems of evaluating a TS were the subject of section 7.4.

8.1 Evaluation Procedures

In principle, three *types of evaluation* can be distinguished. They are characterized by the questions shown in figure 8.1.

Type	Given	Question
I	The realized technical system	How good is the technical system ?
II	An assigned problem (sum of requirements) and a solution proposal or prototype	Does the TS (or the model) conform to the given requirements ?
III	An assigned problem (sum of requirements) and various solutions that technically conform to the requirements	Which is the best, or the optimal solution (alternative) ?

Fig. 8.1 Types of Evaluation

An evaluation can be performed in two basic ways:

- subjective, by feel;
- objective, based on certain criteria.

Evaluation by feel should not be rejected outright. It should not be merely the result of unfounded opinion, but should be based on long experience, and should consciously be cultivated in a systematic way. Only then can one rely on design feel, an intuitive judgement. This is particularly important when information is lacking, as happens frequently at the beginning of an engineering design or development process.

Both of these ways of evaluating should follow the working methods outlined in figure 8.2. The following basic operations occur in each case:

(a) selecting an overall synthesizing characteristic, whose value is to be reported, and from which a decision is to be reached;
(b) selecting the criteria against which the TS is to be evaluated, and which are to be combined into the characteristic;
(c) evaluating the individual criteria, establishing or determining the measures of its properties;
(d) processing the criteria measures into a synthesizing characteristic value;
(e) comparing (to the requirements, a limiting value, or between values) and deciding.

Evaluations almost always lead to decisions, they are in fact usually the necessary precursors to decisions. Such decisions are the answers to the questions in figure 8.1, and are indicated in figure 8.2 as the concluding step, the decision processes. These decisions should not be merely accepted, but some suitable actions should be taken before the final decision is made.

All evaluations should be independently *verified* to ensure as far as possible that no errors have been made, and that the basis for a decision is fully agreed. Any risks (of failure of the proposed TS, the variability of the assessed values, etc.) should be established in as objective a way as possible.

Part of the verifying procedures should be used to check for possible *improvements* which could increase the value of the synthesizing characteristic. One of the possible procedures that can result in improvement is *optimizing*, as outlined in section 7.4, part (6).

These procedures are normally performed by designers, and also by independent groups of engineers during value engineering [22, 84, 94, 144, 158], design reviews or design audits [48] performed to *substantiate* the decisions.

This double or multiple checking is needed to ensure that the most appropriate decision is taken, and is an integral part of the general procedural model of design (refer [119] figure 13). There is a distinct danger that a possible solution may be prematurely rejected on the basis of "feel", sometimes coupled with prejudice, political discomfort, manipulative preemption of decisions, and other destructive techniques (compare [25, 26]). A systematic procedure of evaluation and verification can help to overcome some of these problems.

The set of problems indicated by these activities, and some of the terminology, is discussed in the following paragraphs.

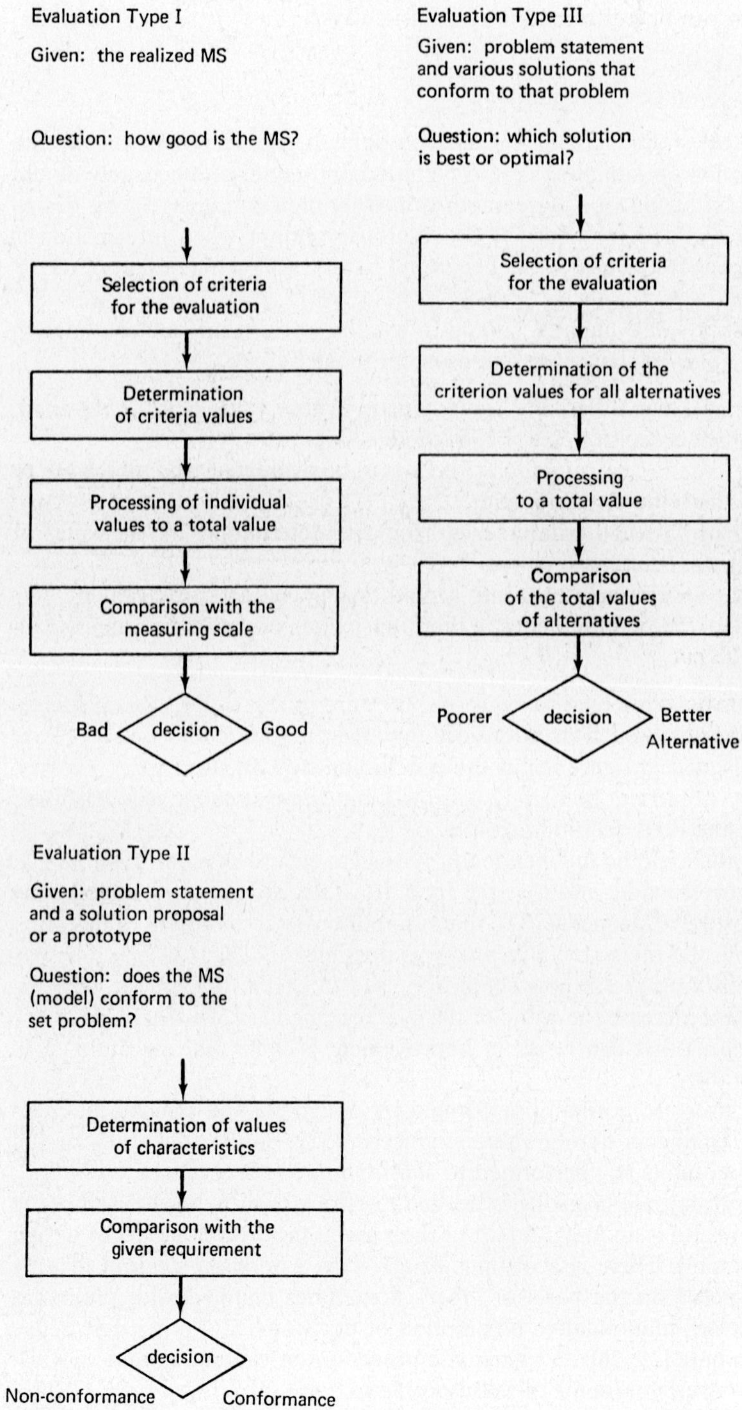

Fig. 8.2 Algorithms of Evaluation (Reproduced from V. Hubka, Principles of Engineering Design, London: Butterworths, 1982 [119] by kind permission of the publishers)

8.1.1 Objective Evaluation

A suitable overall synthesizing characteristic value must be selected. This assumes that the technical system is characterized by certain measures.

By means of the "value" of a technical system, someone's need will be satisfied, or comfort or pleasure aroused. A number of different values can be defined on this basis, using the groupings of properties listed in figure 7.1:

- The *technical value* is a combination (as a vector) of the qualities of technical properties of a given TS. The relevant properties are those grouped in the functions, the functionally determined properties, and the operational properties. One could also include the manufacturing and design properties, but this would be a very comprehensive combination.
- The *economic value* is a combination (vector) of qualities and measures of the economic properties.
- In similar ways one can define both the *ergonomic* and the *aesthetic* values, as well as the transportation, production and other values derived from the classes in figure 7.1.
- The *usage value* combines all usage properties that serve to satisfy human needs. This group consists of the external properties, compare figure 7.15.
- The *total value* can be regarded as the vector resultant of all values, or of the measures of all classes of properties for the given product.

These characteristic values can be termed "absolute", and are based on the concept of an "ideal solution" as carrier of all ideal properties, both with respect to each individual property, and the whole product and all its properties (compare [162]). The ratio of the actually achieved value to the value of the ideal solution can be used as a parameter for expressing the synthesizing characteristic, the relative technical value.

Various other characteristics can also be used, for instance the *effectiveness* as defined in sections 4.7 and 7.2.

In addition, a number of relative characteristics can be defined. These may be objective or subjective. A particular example that is used as a basis for marketing is defined as follows:

- The *esteem value* is a combination of properties that affect people's desire to own the product [84]. It relates particularly to the operational, ergonomic and aesthetic values, but may also be influenced by societal and economic factors. Esteem value is largely reflected in subjective assessments, and results from intangibles such as styling and fashion, together with peer group interactions that make particular brand names or styles attractive as the "in thing". The "law of supply and demand" also operates here.

Kesselring [132] shows how the relative technical value can be contrasted to the relative economic value by means of a two-dimensional "relative strength diagram". Each technical system is characterized by a point in the Cartesian coordinates, as illustrated in figure 8.3. For this purpose, each of these two values must be synthesized into single indicator values expressed as a proportion of the value of the ideal solution. A three-dimensional graph would be possible, using for

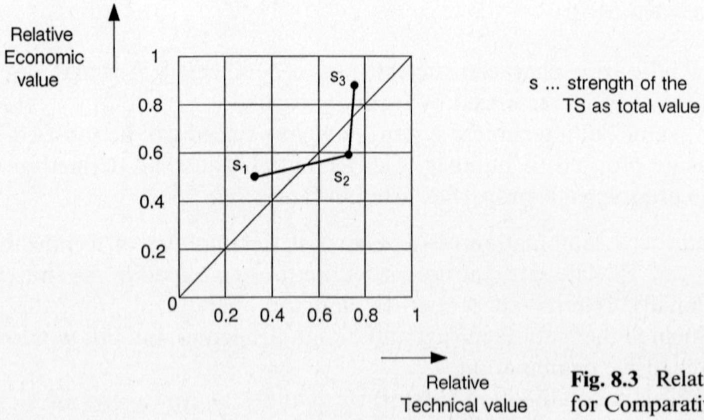

Fig. 8.3 Relative Strength Diagram for Comparative Evaluation

example the relative usage value, but more than three such dimensions cannot be graphically represented.

8.1.2 Selection of Criteria

Selecting the criteria is an important activity for evaluating a technical system, it involves choosing suitable properties of the TS. As a general rule, this selection should be holistic, taking in all aspects of the TS, so that the evaluation will consider all relevant properties and their relationships (see figures 7.1, 7.15 and 7.16). In many cases, for pragmatic reasons, this rule is replaced by the principle of limiting the selection to the most significant ones. When a larger number of characteristics is available and used, a general overview is usually more difficult to achieve.

Criteria must generally be selected according to the significance of each individual property, and with due regard to how easily each property is quantifiable. Selection of criteria is also conditioned by the purpose and aims of the evaluation, and by the phase of life of the TS (the progression from idea to hardware). The situation for an evaluation is very different if the TS exists only as a "black box" process, as a conceptual idea, or as a prototype. In addition, a certain amount of freedom is available in the choice of criteria for evaluation types I and III (figure 8.1), but in type II the selection is determined by the requirements.

8.1.3 Unifying the Criteria

Even though many properties to be used as criteria are quantifiable, combining them into a synthesizing characteristic is difficult because the measures may be expressed in different units. Some possiblities of obtaining a suitable synthesis are (a) to express all criteria in terms of money (compare [84]), (b) to use a form of point-rating (see also next paragraph), (c) to search for a combination of criteria that reveals a significant trend or insight (compare figure 11.4), or (d) to compare each alternative with every other one using only marks to show "better", "worse" or "equal" [176].

In *point-rating*, each property is assessed by giving grade-points, for instance according to the following scale:

Approach to the ideal realization

excellent (ideal)	5 points
very good	4 points
good	3 points
adequate	2 points
only just acceptable	1 point
unacceptable	0 points

This implies that all properties are uniformly evaluated, which is important for further processing of the information. It must be emphasized that for the important and critical properties an evaluation of "unacceptable" is not permitted, such a solution is thus a candidate for unconditional rejection. Only alternatives that satisfy all the requirements are permissible. An "unacceptable" rating with respect to a property or feature that was given the priority "desired" does not necessarily invalidate the solution.

8.1.4 Combining Criteria Measures

In the subsequent *mathematical processing*, various techniques and algorithms are available, see figure 8.4.

The most usual is the simple calculation of an arithmetic mean (average), but it must be questioned whether this average adequately characterizes the value of the TS. The weighted arithmetic mean may be preferable because it recognizes and considers the relative importance of individual properties. The median or the mode may under certain circumstances be a better statistic to use as synthesizing characteristic. These comparisons between solutions are essentially one-dimensional.

A multi-dimensional comparison is realized by using the geometric mean (absolute or weighted). The evaluating parameter may be thought of as the length of side of an equivalent n-dimensional regular solid (all edges of equal length) which has the same "volume" content as the n-dimensional solid calculated from the point-values of properties as its sides.

A third possibility is to use the vector sum, by vectorially adding the point-values of properties. The resultant radius vector serves as synthesizing characteristic.

8.2 TS Evaluation - Statements

Statement 8.1 The total verdict (or partial verdict) about the quality (or total value, or partial value) of a technical system may be characterized by a (vectorial) sum of measures of selected properties.

Statement 8.2 The measures of properties (as their states of embodiment) should be determined by the most objective measuring scale available (problem of quantifying the properties).

No.	Type	Equation	Geometric interpretation	Comment
1	Absolute arithmetic mean	$\bar{p}_1 = \frac{\sum p_i}{n}$		Simple calculation
2	Relative arithmetic mean	$\bar{p}_2 = \frac{\sum p_i}{n\, p_{max}} = \frac{\bar{p}_1}{p_{max}}$		$\bar{p}_2 \leq 1$, comparison with ideal solution
3	Weighted absolute arithmetic mean	$\bar{p}_3 = \frac{\sum p_i g_i}{\sum g_i}$	Mean length	Importance of properties considered
4	Weighted relative arithmetic mean	$\bar{p}_4 = \frac{\sum p_i g_i}{\sum p_{max} g_i} = \frac{\bar{p}_3}{p_{max}}$		$\bar{p}_4 \leq 1$
5	Absolute geometric mean	$\bar{p}_5 = \sqrt[n]{p_1 \cdot\cdot p_i \cdot\cdot p_n}$		for $p_i = 0$ $\bar{p}_5 = 0$
6	Relative geometric mean	$\bar{p}_6 = \frac{\bar{p}_5}{p_{max}}$	"Side" of n–dimensional regular solid with same volume as solid of sides p_1 to p_n	$\bar{p}_6 \leq 1$
7	Weighted absolute geometric mean	$\bar{p}_7 = \frac{\sqrt[n]{p_1 g_1 \cdots p_n g_n}}{\sqrt[n]{g_1 \cdot\cdot g_i \cdot\cdot g_n}}$		Not significant, g cancels
8	Weighted relative geometric mean	$\bar{p}_8 = \frac{\bar{p}_7}{p_{max}}$		As for 7
9	Absolute "vector"	$\bar{p}_9 = \sqrt{p_1^2 + p_2^2 + .. + p_n^2}$		
10	Relative "vector"	$\bar{p}_{10} = \frac{\bar{p}_9}{p_{max}}$	"Diagonal" of n–dimensional solid of sides p_1 to p_n	$\bar{p}_{10} \leq 1$
11	Weighted absolute "vector"	$\bar{p}_{11} = \sqrt{(p_1 g_1)^2 + \ldots + (p_n g_n)^2}$		
12	Weighted relative "vector"	$\bar{p}_{12} = \frac{\bar{p}_{11}}{p_{max}}$		$\bar{p}_{12} \leq 1$

LEGEND :

- p_1 to p_n — Point value of criterion 1 to n
- p_{max} — Point value for ideal solution
- $g_1 \cdot\cdot g_n$ — Weighting factor for criterion 1 to n
- $\bar{p}$ — Mean value
- n — Number of evaluation criteria
- $\sum p_i = \sum_{i=1}^{i=n} p_i$ — Sum of point values 1 to n

Fig. 8.4 Mathematical Processing of Point Values to a Total Value

Statement 8.3 The selected properties are termed criteria; selecting them is one of the most important factors for the quality of the evaluation (or comparison).

Statement 8.4 Processing of the measures of individual criteria to a total value can be accomplished by various methods. The selected method is also one of the important factors for the quality of the evaluation.

Statement 8.5 A correct determination of the measures of properties in some phases of the life of a TS is a difficult problem. An objective determination or estimation of the values supports the quality of the evaluation.

9 Representation of Technical Systems

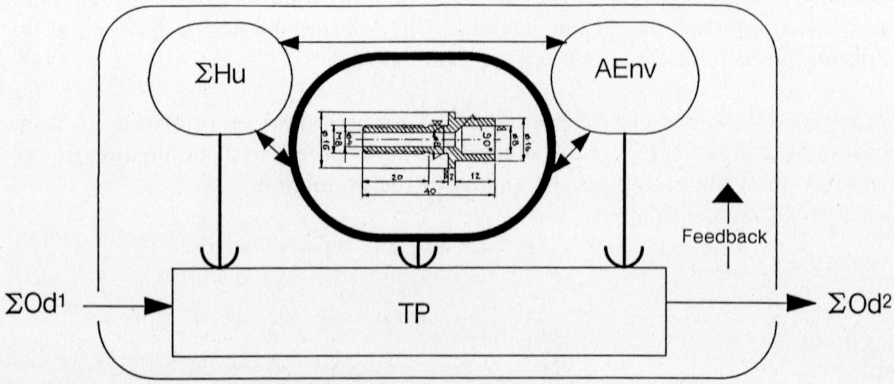

Technical systems are *represented* by various models that describe the design properties of a TS, and thereby also describe the object system itself [111, 112]. Many different kinds and techniques of representation and modelling exist - answering the question "with what?" The most important factors in selecting an appropriate representation or model are (a) the object ("what?"), its state of existence (compare figure 5.2 - "when, where, how, who?"), and the purpose of the model (communication, information, experiment, calculation, simulation, fixation in recallable memory - "why?").

Preparing, producing, using and reading such models is frequently time-consuming, therefore much effort and attention has been focussed on rationalizing these aspects. A revolution has been initiated in this field by the advent of digital computers and data processing techniques. Some problems have been eased, but others have appeared, e.g. modelling for the digital computer, and these are discussed in section 9.2.

9.1 Types of Representation for Technical Systems

Technical systems have been analyzed in the previous chapters, various abstract models have been described (section 5.2), and their properties explained (chapter 7). A few different representations of technical systems were introduced in that context which serve to demonstrate the multiplicity of possible types of representation, particularly in relationship to the degree of abstraction of the TS. Let us now attempt to place these models into an order, and correlate them with the known representational documents. The criteria for this ordering are (a) the degree of abstraction, and (b) the degree of completeness with which the model describes the realized TS (compare figure 5.11).

Figure 9.1 lists the sequence of design documents (from the list of requirements, to the assembly drawing) in the central column, and indicates some of the graphical representations that are appropriate for products of mechanical engineering by arrows from the items in this list. The upper right-hand section of this diagram shows the information content of the relevant document. It indicates which design characteristics are typically established in connection with the design documents, and shows by the hatched fields how information about the design properties progressively increases. For instance, the function structure only makes limited statements about the design properties, and (with the exception of some existing details about action sites) says almost nothing about forms and tolerances of components. In contrast, the detail drawings must contain full information about all design properties of the constructional elements, and the assembly drawing must completely specify their arrangement and structure.

We thus meet the terms "design characteristics" and "design properties" from a different direction, and thereby show definitively that they are a means of describing technical systems.

The types of representation shown in this context by no means cover the full range of possibilities. It is not intended to provide a full coverage of this range of problems, but a few additional techniques can usefully be discussed, especially those that gain importance with application of computers.

In the first instance, an alternative representation of structures (compare figure 5.4) is available with the help of bond-graphs. Usually, each node (junction point) represents an element, and each link (arrow) represents a relationship. A link between two points (a two-place relationship) can be directed (a single-headed arrow) or undirected (a double-headed arrow).

Graphs of this kind can be axiomatically described as follows:

- A graph consists of a set of elements $\mathbf{e}_1$ - $\mathbf{e}_2$.
- On this set, a two-place relationship **R** is defined that may be characterized by **M** using the cross-product of two sets.
- The elements of **M** are represented as nodes, and are graphically shown by various symbols (points, circles).
- The pairs of elements determined by the relationship **R** are represented as links, i.e. as new connections between the nodes.

Such a graph permits a mathematical formulation that can be processed by a digital computer. An example is shown in figure 9.2, a sub-assembly of gear wheel and shaft [213].

A different convention is used in critical path diagrams for production planning (CPM, PERT or similar methods) [47, 96, 149, 197], where each node represents a relationship (an event), and each link represents an element (a task).

Representation can also be accomplished by a coupling matrix which describes the elements by means of the connections between their functions. It is a form of representation that is also suited to computer applications. By emphasizing the relationships of functions, the geometry is naturally relegated to the background. A transformation between bond-graphs and matrix representations should be possible.

Fig. 9.1 Types of Representation of Technical Systems and their Information Content (Composite from [112, 119, 209]) (Heading figure, top right, reproduced from V. Hubka, Principles of Engineering Design, London: Butterworths, 1982 [119] by kind permission of the publishers)

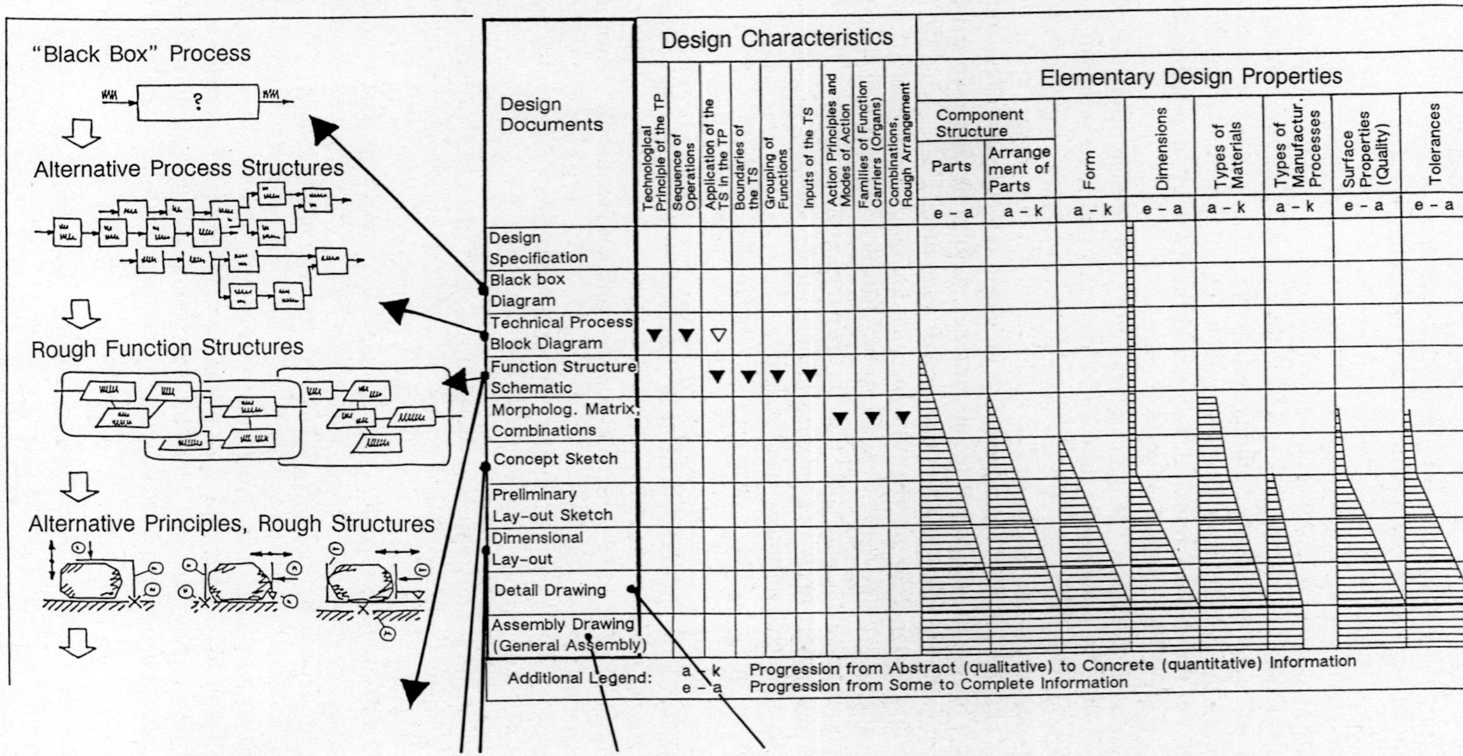

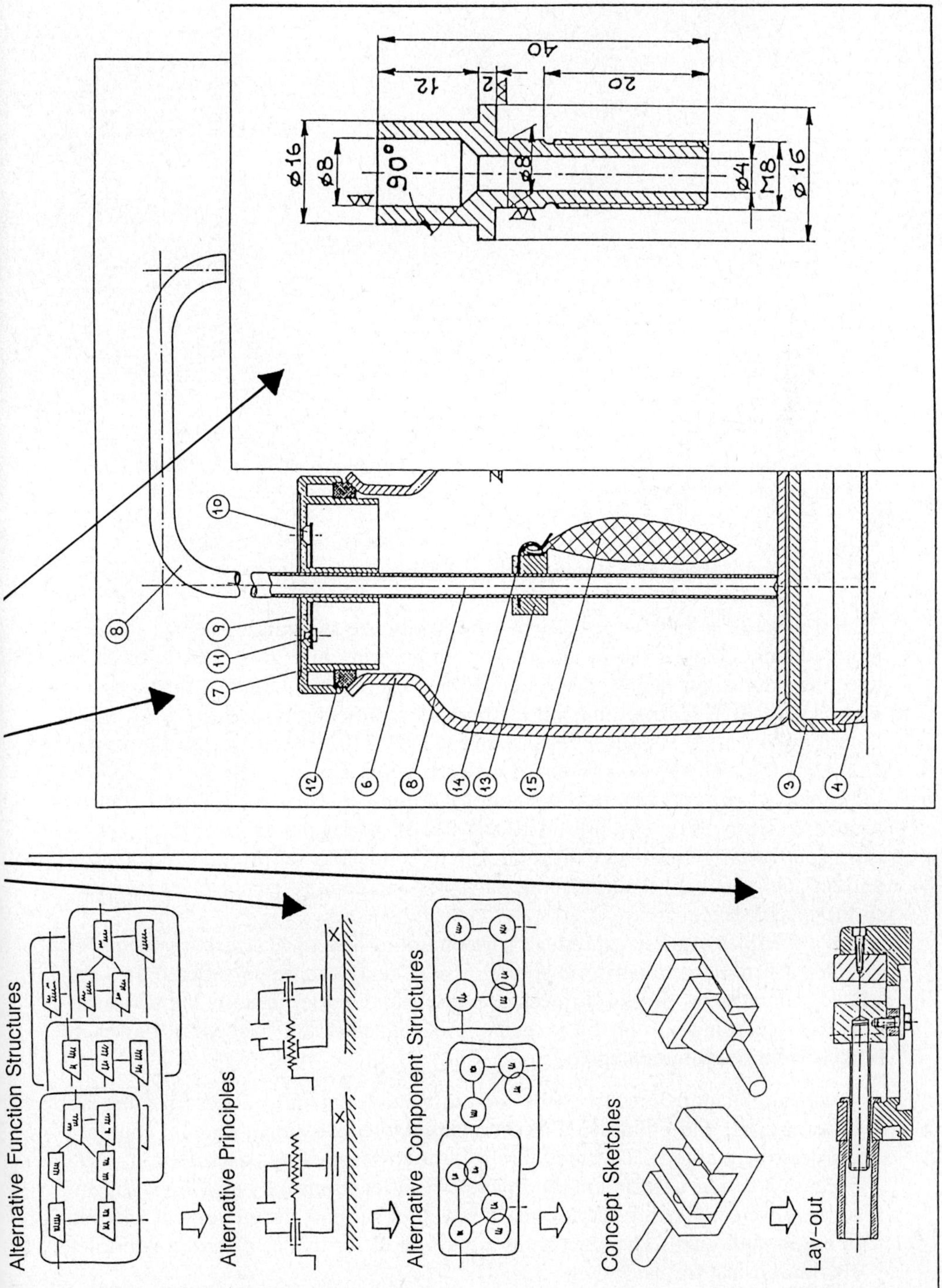

Alternative Function Structures
Alternative Principles
Alternative Component Structures
Concept Sketches
Lay-out

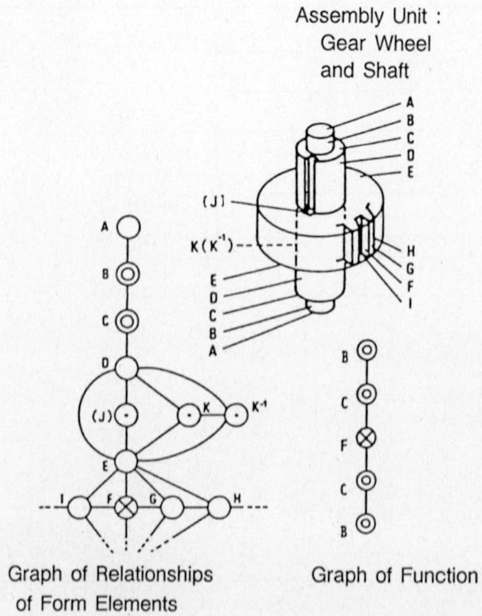

Fig. 9.2 Representation of the Structure of Technical Systems with Graphs [213]

9.2 Representation of Constructional Elements

Technical systems of lowest level of complexity are the constructional elements (components). These exist in all areas of engineering in a great variety of forms and dimensions, especially in mechanical engineering the number of shapes and sizes is very large. Current manual methods of representation are well known and described in the literature (see for instance [57, 82, 87]). The questions of how such components can best be represented with the help of digital computers has been the subject of research and development for a number of years, and improvements occur frequently. Problems arise with transfer of data about geometries and structures, particularly from one digital form of modelling or program package to another, but also between modelling packages and computer-controlled manufacturing machinery.

The principles of these computer representations lead once again to systems terminology. From a geometric viewpoint, a solid body can be considered as a system of points, edges (lines), surfaces or partial solid bodies, usually known as primitives. According to the basic geometric elements, the following classes of models have been implemented:

- A wire-frame model describes the constructional element by edges (or other significant lines) and points, in a direct analogy to conventional manual engineering drawing methods. This model reflects an unsatisfactory description of form, because it can be ambivalent (an image can suffer reversal by an optical illusion of the observer viewing the image), it does not support distinctions of visibility of edges (although some algorithms are available for this task), and it will tolerate nonsensical forms.
- A surface model uses a system of surfaces that bound a solid component. Surfaces provide a unique description of the component.

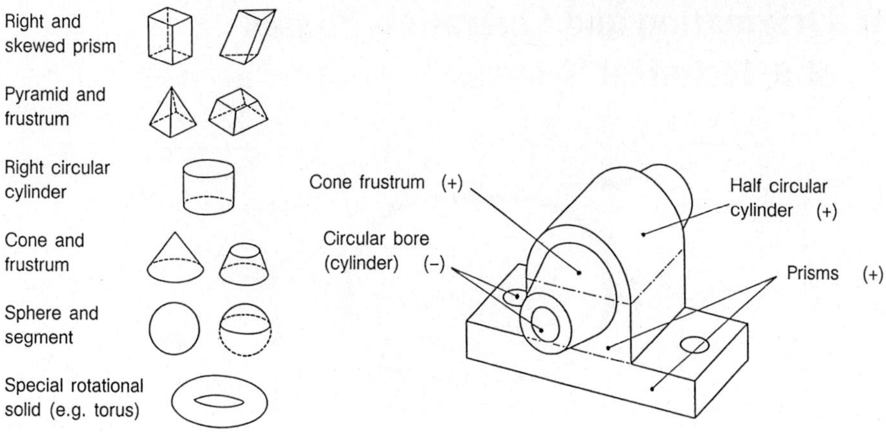

Fig. 9.3 Solid Models of Technical Systems: Basic Bodies (Primitives) and Modelling as Sum of Basic Bodies

- A solid model (volume model) uses a system of partial bodies with spatial relationships. A selection of primitive bodies is usually available, as shown in figure 9.3. This description corresponds accurately with reality, and most forms can be achieved by (Boolean) operations of union (addition), difference (subtraction), and intersection, which are demonstrated by the example in figure 9.3.

9.3 Representation of Technical Systems - Statements

Statement 9.1 Representations of technical systems describe models of TS with their design properties and design characteristics.

Statement 9.2 According to the classes and numbers of the described design properties, representations can be relatively abstract (when abstracting from some design characteristics) or concrete (and usually more complete), compare also figures 9.1 and 5.11. In other words: The model of the technical system described in the representation can be more abstract or more concrete.

Statement 9.3 Every TS model can be represented by various methods and techniques of representation.

Statement 9.4 Every representation serves one (or more) certain purpose, and the kind of representation must be suited to that purpose.

10 Origination and Operation Phases of a Technical Systems *(Ontogenesis and Genetics of TS)*

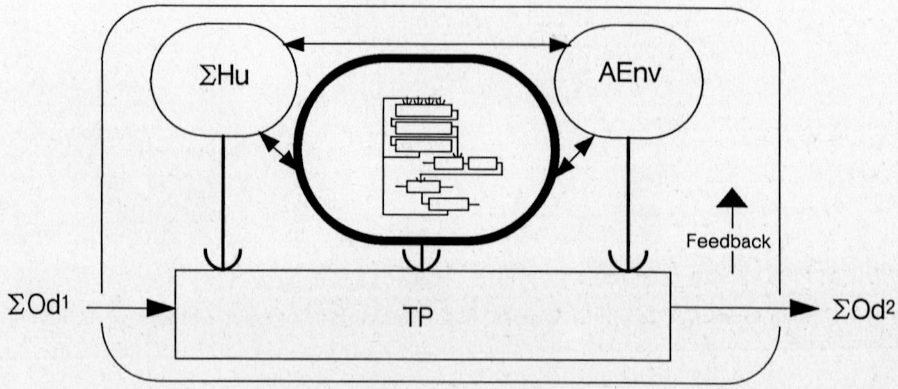

Technical systems are generally very complicated, but their progress from the first idea to the finally realized system, a process that can extend over years, is equally complicated. It is important to recognize all the phases that a technical system must pass through during its origination (and later during operation), not only for planning purposes, but also in order to discover all the factors that will influence the technical system, and that must be considered in preparing and designing it. Designers engage themselves with all phases, even if only at times as consultants.

The whole life-span of a technical system can be divided into four major phases: origination, distribution, operation and liquidation. Each of these phases contains a number of partial processes, divisible down to operations. The contents, relationships and sequences of progress of a TS are the subjects of this chapter.

The basis for this discussion is presented in a block diagram, figure 10.1, which shows the characteristic phases as processes and partial processes (compare also figure 5.2). This diagram and the symbols used are derived from the process model, see figures 4.1 and 4.6, and are defined in appendix B. The processes are named as they apply to a machine system (as products of mechanical engineering) as an example for the life of a technical system; equivalent processes in other fields may have different designations. The terms selected for two of the major technical processes (manufacture of the TS, and working process) reflect particularly the choice of a mechanical device.

For a more extensive discussion of the origination and operation phases, we will also consider the influences of degree of complexity of the TS (see figure 6.2), and number of TS to be manufactured. These considerations lead to formulating three characteristic classes of technical systems:

(1) Technical systems up to the third degree of complexity manufactured in large numbers;
(2) Technical systems up to the third degree of complexity manufactured as single items;
(3) Technical systems of the fourth degree of complexity (plant).

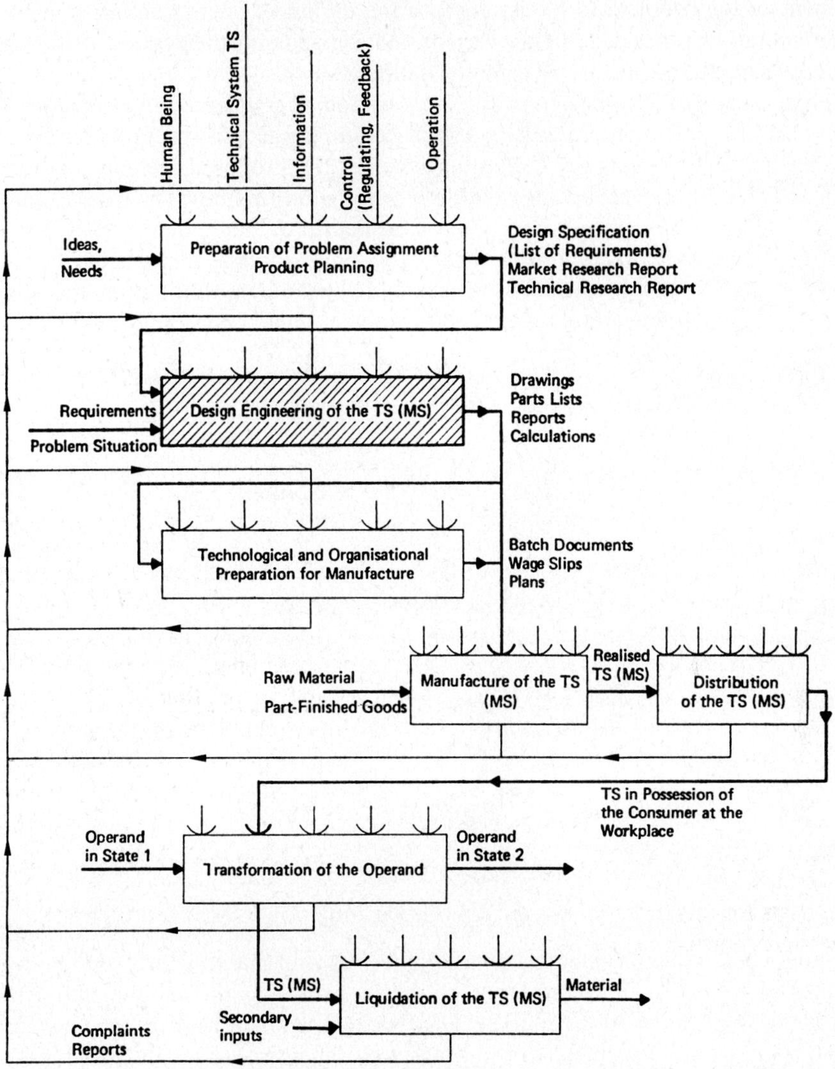

Fig. 10.1 Complete Model of the Origination and Operation Phases of Technical Systems (Reproduced from V. Hubka, Principles of Engineering Design, London: Butterworths, 1982 [119] by kind permission of the publishers)

In order to prepare this discussion, it is useful to indicate general characteristics of the individual processes and couplings in figure 10.1. Defining more exact limits to the contents and applications of these processes would be difficult because they overlap in their work.

The process of *preparation* should reveal that the proposed technical system would, in principle, be purposeful (for the manufacturer, for the market, for society in general or a part of society, etc.) and could be manufactured within limitations of current technologies – a feasibility study. It should also prepare the statement of the problem to be assigned to the designers for the design process. This preparation process is highly varied, and depends on the type of technical system and requirements, as indicated in figure 7.21.

Research is an important part of this preparation. It is included in the preparation process because some research work of certain kinds must precede design to deliver a part of the basis for evaluating whether the proposed technical system can be realized. Different tasks for research range from finding basic insights and theories about natural phenomena – basic, scientific or fundamental research – and obtaining technically applicable data – applied research. The linkage between basic research and new products is indirect, and generally does not affect the processes of originating new products, except in some isolated cases. Figure 10.2 indi-

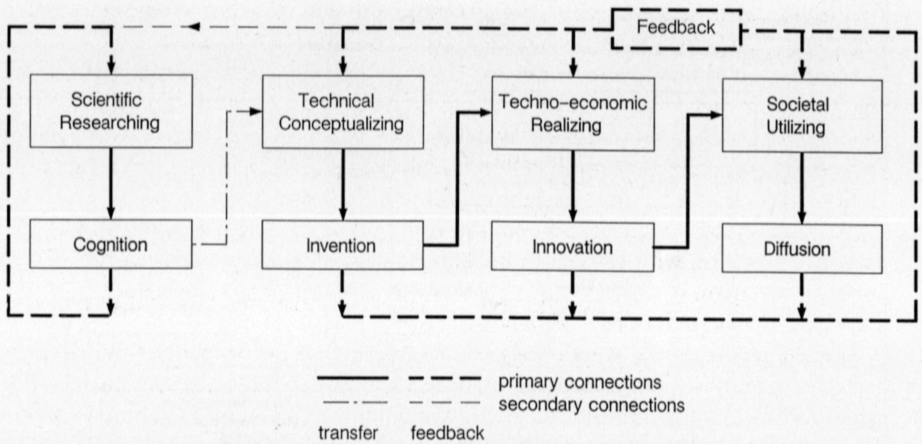

Fig. 10.2 Phases of Technical Ontogenesis – Cycles of Research and Technical Development [65]

cates how basic research interacts with engineering and manufacture, and shows the roles of invention in creating potential new products, and of innovation in bringing these products to the market. The feedback loops lead from the customers and users to the designers, and into research. The discussions in this chapter are concerned with the industrial sphere, we will therefore refrain from further discussions of basic research.

Goal-oriented applied research is essential for every organization that wishes to improve its technical level, or maintain its level relative to that of competing or cooperating organizations. Various types of applied research can occur in an organization:

- Research with respect to the transformation in a working process. For instance, some of the relevant research in machine tool construction should be concerned with metal cutting processes, and a manufacturer of food processing equipment should conduct research into the processes for food production.
- Research to achieve better properties of the technical system. The search may concern appropriate means, including working principles, structures (at various

levels of abstraction, see figure 5.4), materials, forms, surfaces, life-span, and so on.

- Research on the market. Investigations cover acceptability of existing products, improvements that are perceived to be needed, competitive products, and market gaps where new products could be aimed.

Applied research uses (among other knowledge) the practical experience of the organization and the insights of basic research. Experiments are an important part of this process. In the course of such applied research, experiments on small-scale processes and operation of pilot plant are often performed.

The output of the preparation process is the definition of the problem to be assigned, i.e. the sum of the requirements for a technical system.

The process of *designing* comprises all the operations of design work, from the first ideas and lay-out sketches to the completion of all detail and assembly drawings, parts lists, etc. that represent the output of the design process.

Further experiments may be needed during or after the main process of designing, frequently on prototypes or working models, to accomplish the separate task of *development* of a TS. This task is defined as achieving the necessary properties (especially sufficient performance) on a realized technical system to overcome the deficiencies remaining after designing and manufacturing [8].

The completed design documentation is the input for the processes of *manufacturing preparation*, with respect to both technical and organizational aspects. This includes selection of manufacturing processes, jigs, tools, fixtures, time plans and prescriptions, material ordering, allocating process plant, and so on.

In the *manufacturing process*, or production, the TS itself is made, and is therefore the operand of the process. The input materials are transformed into the finished technical system, based on the information delivered by the preceding processes (as one of the operators), by effects delivered by human hand and suitable process equipment (e.g. machine tools). The output is the finished TS, and represents the end of the *origination* phase of a technical system.

The purpose of the *distribution process* is to change two properties of the technical system, namely ownership (from manufacturer to user) and location (from manufacturer's factory to usage location). Typical partial processes are storing, advertising, packaging, transporting, erection, assembly, installation on site, etc. (compare [119] figure 7).

The technical system is then used in the *working process*. The TS acts as operator and exerts the necessary effects onto the operand to perform the desired transformation. Together with the other operators (humans, active environment, etc.) it transforms the properties of the operands from their input values to the desired output values. This transformation is the main aim of the whole activity. During operation, the technical system must also be serviced and maintained, and if necessary modernized.

After completing the duration of use (after life is ended, see also section 7.2.7.3), the technical system again becomes the operand, and is transformed by the *liquidation process* into waste and scrap. Scrap can be recovered, and re-cycled into renewed material resources for manufacture (see also section 7.2.7.12).

10.1 Origination Phases of Technical Systems Manufactured in Quantity

The origination and operation phases of technical systems manufactured in large quantities are listed and characterized in figure 10.3, and shown in the block diagram of figure 10.4 derived from the list. The first phases follow the general pattern outlined in the introduction to this chapter, this discussion therefore starts with conception and lay-out.

Phase	Important Partial Processes in Phase	Important Background	Results (Output) of Phase
Basic Research	Processing of scientific information, analyzing, formulating hypotheses, model building, experimenting, reporting	Scientific information, stimulation from practice	Research reports, scientific papers, fundamental data
Applied Research	Processing of scientific information, analyzing problems and tasks, formulating hypotheses, model building, experimenting	Scientific & technical information, technical problem formulation	Research reports, test reports
Market Research	Investigating home and foreign markets, competitor situations & perspectives on economic developments	Statistical investigations, offers, brochures, reports	Market research reports, sales prospects
Development Decisions	Studying feasibility, processing information, evaluating & deciding on technical & economic criteria	Reports on company policy, patent search	Report on decision with substantiation
Concep–tualizing	Defining the task & problem precisely, establishing process & technology, establishing functions & means, calculating & evaluating	Problem definition, techn./econ. information, research results, patent reports	List of requirements, calculations, preliminary lay–outs, reports
Laying out	Establishing means, arranging, establishing form, calculating, dimensioning, techn./econ. evaluating, tolerancing, drawing	Results of conceptualizing, drawings of existing TS, techn./econ. information	Calculations, dimensional lay–outs, reports
Detailing	Establishing form of machine elements, calculating, establishing materials,dimensioning, tolerancing, drawing	Lay–out drawing, standards, material lists, lists of manu–facturing equip.	Complete production documentation and instructions
Prototype Preparation	Establishing manufacturing technology, estimating time to design & make jigs & tools, ordering materials, planning manufacture	Production documents, material & work standards	Batch documents, wage slips, plans

Fig. 10.3 Origination and Operation Phases of Technical Systems in Quantity Manufacture

Phase	Important Partial Processes in Phase	Important Background	Results (Output) of Phase
Prototype Manufacture	Preparing materials, operations of manufac-turing, assembling	Production documents, preparation plans	Prototype, report on suitability for manufacturing
Prototype Testing, Developing	Preparing test plans & equipment, obtaining mat-erials & personnel, testing, evaluating, report writing	List of requirements, design documents reports, standards	Test report
Production Decision		All documents	Decision
Correction of Manufac-turing Documents	Lay-out for corrections, altering drawings & parts lists	Production documents, reports	Documents for quantity manufacture
Preparation for Quantity Manufacture	Manufacturing technology, dimensions of raw materi-als, work sequence, time setting, designing & making jigs & tools, organizational preparation : material deliveries, manufacturing plans, arrangement of machine tools	Production documents	Production plans & means, organiza-tional documents
Pre-production "Zero-batch" Manufacture	Preparing materials, machining operations, assembling	Production & planning documents	"Zero-batch" products,report on means of production
Pre-production "Zero-batch" Testing	Preparing for testing, testing, evaluating, reporting	List of requirements, Prototype test	Report on fulfillment of required properties
Correction of Documents	Correcting TeS documents, improving means of manufacturing	Manufacturing documents, test reports	Mature manufac-turing documents & means
Quantity Manufacture	Preparing materials, machining operations, assembling	Manufacturing documents	Quantity-manufactured technical systems
Distribution	Storing, exhibiting, tendering, selling, packing, transporting, maintaining, accepting	Instructions, market reports	Sold TS
Operation of TS	Working, cleaning, adjusting, maintaining, modernizing	Instructions, manufacturing documents, reports on operation	Complaint reports on TS
Liquidation of TS	Disassembling, preparing for recovery or destruction	Instructions, manufacturing documents	Waste, scrap

Fig. 10.3 (cont.)

10.1.1 Conceptualization and Lay-out

The technical system is usually conceptualized, and the concepts embodied, laid out, and detailed by experienced design engineers, or a design team. The systematic processes of designing, and current knowledge about the science of design, in this important stage of developing a TS are outlined in chapter 12 (see also [119]). The documents that usually result from this activity are listed as follows:

- Report about the results (or the state of progress) of the research and other investigations that may have been done before this problem was assigned to the designers.
- Various proposals for abstract structures (compare figure 5.4), including concepts, and their rough embodiments in preliminary lay-outs.
- Design (dimensional) lay-outs of the product and its important assembly groups (in a form close to that of the final assembly drawings).
- Function diagrams and kinematic schematics.
- Reports supporting and explaining the technical and economic viewpoints for

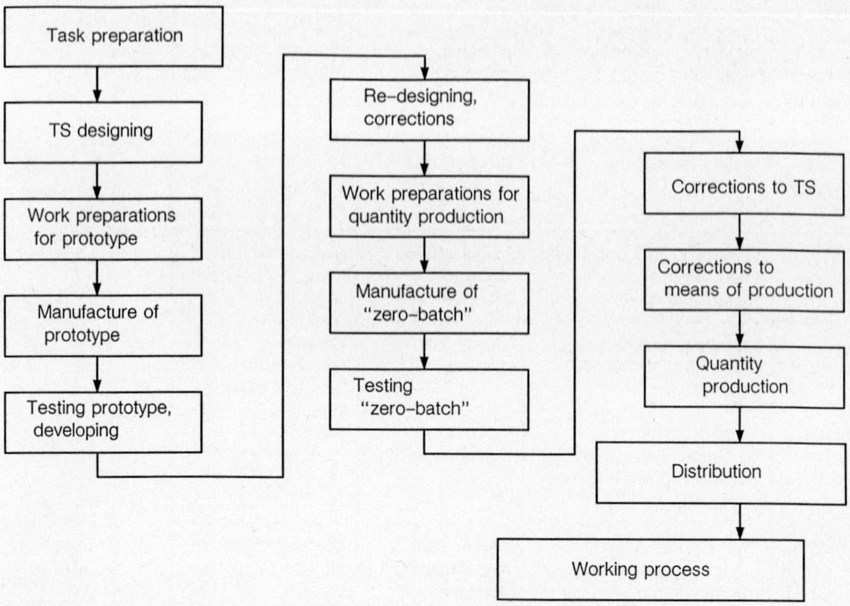

Fig. 10.4 Simplified Block Schematic of Origination and Operation Phases of Technical Systems in Quantity Manufacture

the proposed solution. Sometimes it is necessary to make a preliminary technical assessment on the assigned problem.

- Results of patent searches to show that the laid-out solution does not violate the patent rights of others.
- Indication of how (in which way, and how well) the solution fulfills the requirements. Proposals about possibilities of fulfilling more demanding requirements that can lead to improvements of the product.
- Preliminary economic analysis and evaluation of the results of these design developments.

In view of their importance for further progress, the dimensional design lay-outs and other documents about the proposed TS should be subjected to a design review and audit by a group of experts, and receive approval [48]. In principle, most properties of the future technical system have been decided, and the decisions recorded in the design lay-out. For this reason such documents should receive particularly careful and detailed attention. Careful preparation and verification of the design lay-outs should reduce the work needed in experimental development on the prototype and the finished technical system.

After approval of the design lay-outs, *detailing* the technical system is started. The first realization (for a quantity-produced system) is usually a prototype, and care must be taken that it possesses the same quality as the finished product. In particular, the prototype must conform as closely as possible to the drawings and manufacturing rules with respect to materials, dimensions, tolerances, and surface quality. The *manufacturing preparations* for the prototype must ensure that a minimum number of jigs, fixtures and special tools are needed to manufacture it.

10.1.2 Manufacture of the Prototype

This phase should include verifying all manufacturing properties, particularly with respect to assembly and adjustment operations [101]. As the designers know which of the elements and structures to be manufactured will incur difficulties, they or their representative should also pursue the processes of realizing the TS. All errors and difficulties should be discovered and reported to the design office, so that the manufacturing documents can be corrected in preparation for the initial (pre-production or "zero") batch - a serial number for these TS may contain the number "zero" in the place reserved for references to batch counts or modifications.

10.1.3 Testing the Prototype

This process, usually involving *development* [8, 204], is an important factor for achieving the required properties and their measures in the realized system, and determining the achieved measures and values of the technical system. Testing is also important for collecting experience, particularly for young designers. They can compare their ideas and hypotheses with reality, attain self-assurance and confidence, and profit from mistakes - their own, and those of others.

A prototype should be tested under increasing loads, up to (or sometimes even exceeding) the required maximum load and most difficult working conditions.

This process serves to confirm the functions, and establish measures for the functionally determined properties, and as many of the operational properties as possible. Suitability of the technical system for servicing and maintenance should also be examined.

The sequencing and results of tests must be recorded (preferably in writing or hard-copy) in a systematic order, particularly with respect to any alterations and improvements made on the prototype. This reference to improvements during testing and development must be coupled with a warning. If the prior design work has not been performed with the necessary care, a situation of "running re-design" can easily arise. In those circumstances, the changes in the prototype system are of larger scale, and therefore more costly, mainly because the deficiencies because of inadequate design must also be overcome. The additional costs (and cost overruns) can easily jeopardize the whole project.

The instructions (anticipating the user manuals) about corrosion protection, packaging, transport, installation, commissioning, servicing, maintenance, operating and even some of the delivery documentation should be prepared and checked.

An important process in testing is *evaluating a prototype*, and making decisions about batch manufacture. The experiences gained by preparing, manufacturing and testing the prototype are the basis for this evaluation. The documentation on this evaluation should be supplemented by assessments of the appearance, transport and ergonomic properties, and of conformance to laws, standards, codes of practice, etc. A further patent examination is often necessary. The economic properties can also be determined with sufficient accuracy.

Having brought the analysis of marketing possibilities up to the most recent state (both from the viewpoint of the technical system to be marketed, and of changes in the manufacturer's situation and the market itself during the period since the last review), the final documents exist from which one can decide whether the manufacture of this technical system could lead to economic success. Sometimes it is more sensible to write off the large costs of design and development, and not to produce, than to manufacture an unsuccessful product under the pressures of amortization.

After the prototype has been tested, the final stage of work with the prototype is *correction* and *modification* of the design documentation for the purpose of producing the drawings and documents for the technical system to be functional and manufacturable in batch or mass production conditions. Even though this work tends to be tedious, and is thus unpopular, it is important for the practical and technical experience that it delivers. In order for this feedback to the designers to be fully effective, it is recommended that the errors discovered, and changes made to the prototype, be analyzed to find their causes (see also [149]).

The third origination phase of a technical system is batch or mass production. It runs in a number of sub-phases, of which the first is manufacture of a pre-production (or "zero") batch that constitutes the dress-rehearsal for full-scale production.

10.1.4 Preparing for Manufacture

This phase consists of general work preparation, planning, machine allocation, generating the documents to regulate and control manufacture, etc., but also manufacture or procurement of the necessary tools, jigs, fixtures and special production equipment. This phase is similar in its execution to that of a new technical system. It too needs great attention in organizational respects. This is particularly reflected in the work of the jig and tool designers who are almost always under extreme pressure to meet time deadlines and cost limits.

Manufacture of the pre-production ("zero") batch should above all examine and test the suitability of the production equipment. As the products of this pre-production batch are used, all their properties are again tested.

It is important to remember that during batch or mass production a continuous effort to rationalize production should be made, in order to achieve improvements in the production equipment, in aids to manufacture, and even in the product itself. This is particularly true for assembly, where much of the work is done by hand. Mechanization or automation of this work usually requires certain special properties of the components. For this reason the designers of the technical system and of the jigs and tooling should be in constant touch with the production organization. From the organizational point of view, this problem of continued improvements in the TS and its production can be modelled by *learning curves*, a technique that must also be kept under continuous review [203].

The phases of distributing and operating the TS have already been discussed. Nevertheless, the designer must keep continuous and careful contact with the technical systems in these phases (see the feedbacks in figure 10.1), so that the system will reach and maintain technical maturity during the course of time [77].

10.2 Origination Phases of a Technical System Manufactured in One-Off Production

Technical systems that are manufactured in small quantities, in the limit as just one system (one-off), can exhibit widely different degrees of complexity with respect to design.

One group of such TS is characterized either by possessing new design properties (e.g. structures at various levels of abstraction, materials) or by application of a new working principle to achieve the transformation of operands. Such systems can be regarded as new developments, for which a "prototype" is manufactured that is intended for normal operation after development. This situation makes large demands on the design process; the output - the technical system - "must succeed", the required properties must be achieved. If this aim is not fully attained, the re-working and changes should be as small as possible, and must not alter the conceptual basis.

Full functional tests of the completed TS can in many cases only be performed at the user's site, the manufacturer can in most cases only test partial functions.

A second group, even though they have been manufactured according to the user's requirements, consists of those technical systems that are already known with respect to their functions and other properties, and therefore present almost no risk. The manufacturer must still maintain the usual vigilance, attention and care so that errors are avoided.

The origination of a technical system to be manufactured as a one-off or in small quantities usually takes place in the following major phases:

- Specifications and contract;
- Conceptual design;
- Design lay-out (embodiment);
- Detailing;
- Production preparation, including manufacture of any special tools, jigs and fixtures;
- Manufacture of the technical system;
- Commissioning, operational testing, operation (application).

The manufacturer and the customer usually define all important technical data for design, delivery and testing in a purchase contract. Conceptual design should consist at least of confirming any existing concepts.

Execution of the dimensional *design lay-out* is guided by the character of the product. If it is part of a capital project, the designer must work closely with the project engineer and the process engineer for the plant. Attention must be paid to the production facilities and possibilities of the manufacturer, and components must be designed within these limits.

In order to increase the operational safety of the whole plant, it is useful, if possible, to test the functions of new and technically important parts, and parts delivered from other suppliers, under normal operating conditions. The designers must again produce complete manufacturing documentation, including instructions for packaging, transport, foundations, adjustment procedures and limits, assembly (at the manufacturer's location and on the final operating site), start-up operation, testing, normal operation, emergency procedures, and maintenance and repair. If possible, even final disposal of the TS should be considered, especially if any hazards (to people or the environment) can be recognized (see also section 7.2.7.12).

Testing of these products is performed according to the conditions agreed with the customer. When these tests are to be performed at the manufacturer's plant, the conditions for preparation and execution of testing should be written into the contract. All experiences, errors, suggestions or implemented improvements must be recorded. Where possible these experiences and improvements should be reflected by modifications of the design documentation, so that they may be avoided in a possible repeat-contract for identical plant.

During *operation*, all properties of the technical system should be continuously observed and monitored. The designers should pursue the operations for an extended time, because only by such observation can they determine the functions, life-span and other properties [77]. Many defects are discovered only after a long operational time, for instance premature wear on certain parts, corrosion and erosion, fatigue, etc. Contacts between the designers and the users is beneficial for both parties, according to general experience. Good service is of particular importance, for the designers and for the reputation of the company.

An account should be kept for each delivered technical system, in which all experiences, suggestions and comments about the defects that occurred, and about maintenance and repair work are recorded. Cross-referencing and categorizing this information within and between contracts is also beneficial, so that continuous checking can avoid errors, and can propagate good features from one system to the next. This task can be aided by generating a suitable data-base on a computer (with the help of available programs).

It is generally not recommended to attempt to shorten the origination phases by omitting or curtailing a phase. Reducing the transit time by parallel work is possible and useful.

10.3 Origination Phases of a Technical System of Fourth Degree of Complexity (Plant)

Compared to the previously discussed technical systems, the systems of fourth degree of complexity are a special case. They include such equipment as a transfer line for mass-production, a workshop or a factory. The creators are a team, typically led by the project engineer. The project serves as the mental model of the equipment, which includes all the properties of the technical systems as previously discussed. The components for the project are generally purchased as completed technical systems, or manufactured to contract. The project design process is therefore somewhat different, in that the project engineering team does not generally decompose the functions from their state of finality (see section 5.1) into the elements; the lowest level of discrimination employed in this decomposition is usually the third degree of complexity, and in exceptional cases the second degree. The main activity of designers is creative work towards establishing the structures (see figure 5.4), i.e. selecting suitable means for the functions of the third level of complexity, and establishing their arrangement in the available space. The problems of originating the selected means is then passed to other engineering designers, and takes one of the forms described in the previous two sections.

As the project engineers and designers are dealing with structural elements (usually large-scale components) that are generally arranged at distances from one another, the connecting functions play a large role beside the working functions. The technical systems used as elements, and their couplings, usually belong to various different fields of engineering, mechanical, electrical, electronic, and so on. The project engineers therefore depend on cooperation from other specialists, and their work must be coordinated by means of suitable working methods.

Project work must take place in at least two phases, and must always be connected with a thorough economic evaluation. In most cases, two phases are not enough, particularly when the first conception of the project only allows inaccurate formulation of the requirements for a set of subordinate execution projects (e.g. as contracts placed with suppliers). For instance the project of a factory hall contains execution projects for the building, the machine equipment, foundations for such equipment, connections and transmission equipment for electric power, gas, compressed air, water, waste disposal, etc.

The origination and operation of a *plant* can be divided into the following basic phases, and stages of progress within those phases:

- Project a) Pre-project - a preliminary study in which the task definition is concretized, project ideas are suggested, and basic solutions sketched out; this work serves to enable decisions about possibilities for realizing the systems - feasibility.
  b) General Construction Project - basic conception as the main study for the project; survey, main and arrangement plans which also contain the contact points for all partial or detail projects.
  c) Detail Project - detail studies of execution plans for all plant and equipment, and coordination plans.
- Execution Usually this involves ordering and contracting for equipment and its assembly based on the lists of requirements, lists of materials and parts, and detail plans.
- Trials Operation similar to testing a prototype of a technical system which should verify the properties of the system and plant.
- Operation Exploiting the plant is the sum of working operations of the individual technical systems, the comments about operation phases as described in sections 10.1 and 10.2 are also valid here.

10.4 Origination and Operation of Technical Systems - Time Dependency

In the preceding sections of this chapter, the partial problems of the phases of origination and operation of technical systems have been described with respect to their contents and mutual relationships. This section deals with relationships of time, in order to show the factors that influence the progress of these processes.

Immediately we discover a contradiction. Both the customer and the manufacturer desire the shortest possible origination time, because these create favorable conditions for low costs and rapid delivery of the product. The customer also wishes to have the best possible quality of the product, but this is connected with a longer time-span for designing, developing and manufacturing, and greater costs. This relationship was shown in figure 7.23 with respect to design work. A similar situation exists in preparation for manufacture, and manufacture itself.

Let us compare the time relationships of a technical system that is produced in quantity (see figure 10.5) with one that is manufactured as a one-off (see figure 10.6). For the system that is produced as a one-off, the number of processes as recorded is smaller, and the care of preparation and detail shown in the diagram reflects the different needs for monitoring the processes in these different production situations. The manufacturing period is relatively larger for one-off production (assuming equal total scope of both projects), mainly because special jigs and tools are avoided, and many manufacturing operations must be performed on universal machine tools by personnel with higher skill levels. In particular, the processes of assembly strongly influence the manufacturing duration because many of the operations must be done by hand in time-consuming ways through lack of suitable assembly fixtures and experience. Possibilities for mechanization or automation present themselves mainly for TS (or groups of similar TS) produced in sufficient quantities

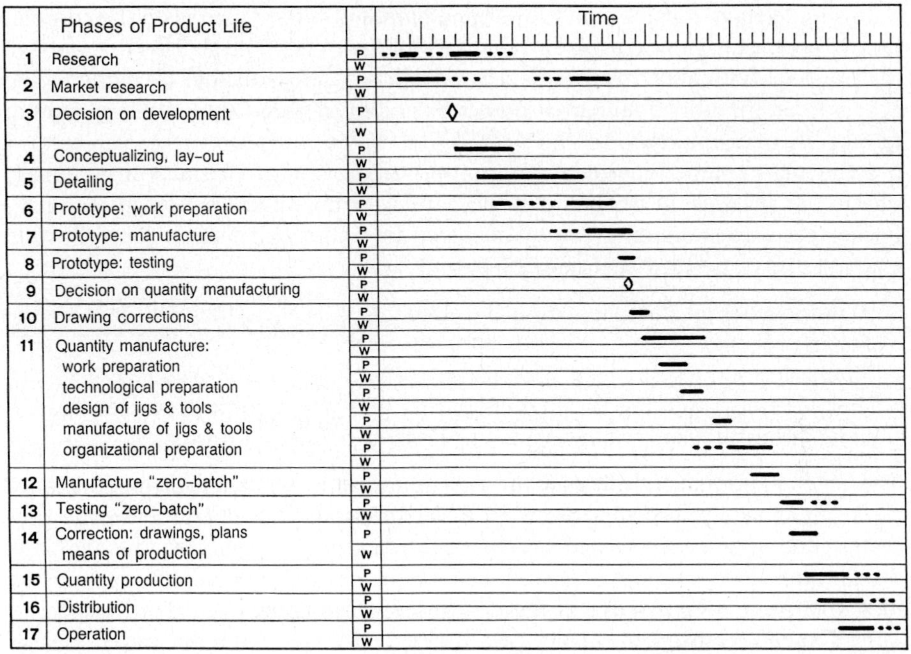

Fig. 10.5 Model of Elapsing Time of Technical System in Quantity Production

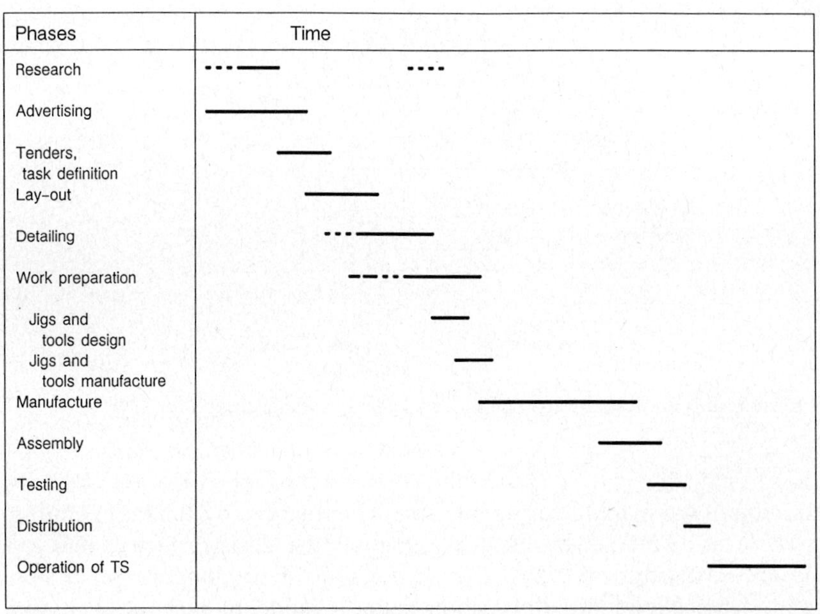

Fig. 10.6 Model of Elapsing Time of Technical System in One-off Manufacture

It is understandable that the start of one partial process depends on the completion of other partial processes, and that this increases the time needed to complete the project. Shortening this time as desired can be achieved either by reducing the time needed for individual partial processes and operations, or by running them in parallel by overlapping.

Reducing the time required for partial processes is most effective in the longer phases, for instance in designing, in preparing for manufacture, and in manufacturing. Extreme shortening can be achieved by omitting a phase. The following examples serve to show such possibilities:

- Acquiring design documentation by obtaining a licence, thereby omitting the design process;
- Purchasing some systems of finished parts (modules or complete technical systems), thereby omitting the manufacturing and preparation processes, and in some instances also certain sections or tasks of the design process.

The general factors that influence the duration of processes are described in chapter 3; these are in particular the working methods and the operators of the processes. For instance, in design processes the duration depends on the available design capacity (see figure 7.23), and the time for solving a typical task depends on the expertise and knowledge of the design team, see figure 10.7. With respect to improvements in the performance of a design team, this figure illustrates a learning curve. The influences of other operators on the duration of the design process is known, at least in broad outline.

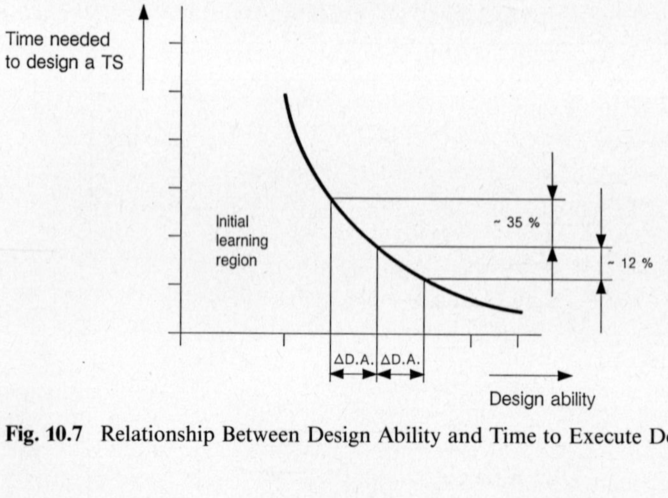

Fig. 10.7 Relationship Between Design Ability and Time to Execute Design for a TS

The duration of the manufacturing processes is influenced by similar operators. For a given level of manufacturing technology as the decisive factor, one can determine an ideal (shortest) time. This is the sum of the longest consecutive sequence of technically required operation times. A model of such an ideal time process is illustrated in figure 10.8. The real duration is frequently a multiple of

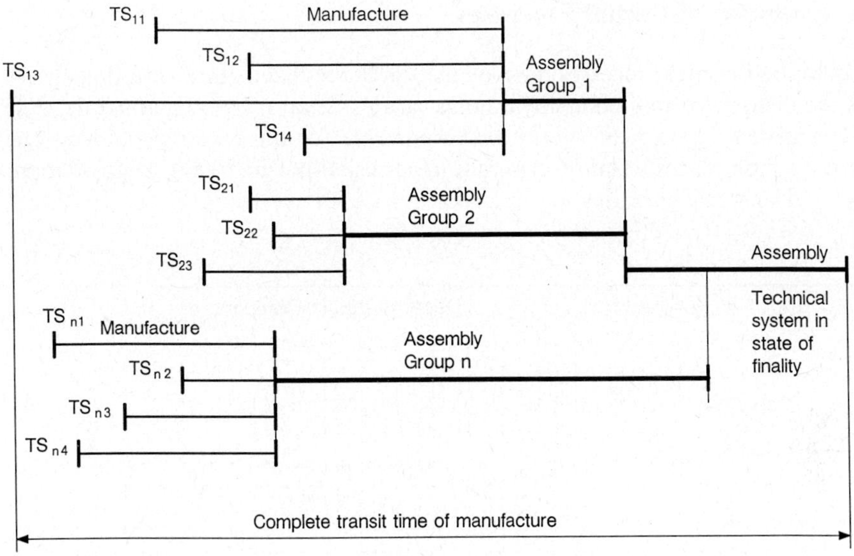

Fig. 10.8 Ideal Transit Time for Manufacture of a Technical System with Given Manufacturing Technology

this ideal time, and depends critically on the management of the manufacturing operations. Planning for shortest durations, and management of the resulting project, is assisted by various techniques such as Gantt charts (similar to figures 10.5 and 10.6), critical path networks, and PERT (Program Evaluation Review Technique) [47, 96, 149, 197].

For the remaining partial processes, the influences of the operators on the duration of the process must be determined.

Overlapping of partial processes is possible if it is permitted by the mutual relationships between the processes and the availability of resources. Parallel work often depends on the designers releasing documentation that is incomplete or not definitive, which can result in negative effects on further work. The earlier manufacturing processes are bound by the released documentation, and the later steps of designing are constrained by lack of freedom to search for better solutions. Alternatively the work is adversely influenced by changes in the design documentation, which usually increases the duration of the process, and its cost.

The useful life of a technical system, the duration of its full capability for operating, depends on wear and other deterioration, and on newer, more modern and competitive products appearing on the market. The latter case implies that the functions or the functionally determined properties are obsolescent, and the product is no longer "up to date" (see also criteria for life ended, section 7.2.7.3). From the point of view of the user it would be advantageous if the technical system could work as long as possible, in order to distribute the amortization costs over a longer period. If a technical system is obsolete, it has been sufficiently overtaken by a new one in the user's assessment, then it should in general be withdrawn from service.

10.5 Execution of Partial Processes

The described partial processes must be executed, regulated and controlled. These tasks are distributed in a company among various departments. In principle, only one designated department should be responsible for the execution of any one task. A possible distribution of tasks and responsibilities between various departments is shown in figure 10.9.

	Phases of Product Life	Departments in Company													
		1	2	3	4	5	6	7	8	9	10	11	12	13	
		Production planning	Research	Development	Design	Sales, service	Work preparation	Prototype workshop	Testing	Jig and tool design	Manufacture	Procurement	Toolmaking	Management	User, customer
0	Product planning	●		○											X
1	Research	○	●	○											
2	Market research	○	○	○		●									
3	Decisions on development	○	○	○		○						○		●	X
4	Conceptualizing, lay-out		○	●	○										
5	Detailing				●										
6	Prototype: work preparation			○			●								
7	Prototype: manufacture			○	○		○	●							
8	Prototype: testing		○	○	○			○	●						
9	Decision on quantity mfg.	○		○	○	○	○		○		○	○		●	
10	Drawing corrections			○	●										
11	Quantity manufacture:														
	work preparation				○		●								
	technological preparation				○		●								
	design of jigs & tools				○		○			●					
	manufacture of jigs & tools									○			●		
	organizational preparation	●										○			
12	Manufacture "zero-batch"						○				●				
13	Testing "zero-batch"			○	○				●						
14	Correction: drawings, plans			○	●										
	means of production									●			●		
15	Quantity production				○		○				●				
16	Distribution			○	○	●									
17	Commissioning				○				●		○				
18	Trial operation			○	○				●		○				
19	Normal operation				○	●									X
20	Maintenance				○	●									X
21	Modernizing				○	●									X

Legend
● Department responsible for execution
○ Cooperating departments
X Possible participation of user

Fig. 10.9 Work Distribution in the Origination and Operation Phases of a Technical System

10.6 Origination of Technical Systems - Statements

Statement 10.1 The life of a technical system is a process system that can be defined as a system of partial processes and their relationships.

Statement 10.2 The four characteristic basic phases of a TS life are: origination, distribution, working operation, and liquidation (see figure 10.1).

Statement 10.3 Every technical system must pass through all these phases, and fulfill the requirements of each phase, i.e. it must possess appropriate properties (see statement 7.1).

Statement 10.4 The number and scope of the smaller stages within the four basic phases depends on the following factors: complexity of the technical system (figure 6.2), originality of design (section 6.7), type of production (section 6.8), and customer (see figure 7.21). According to the type and traditions of the organization executing the phases, they can be very varied in their content and conformation.

Statement 10.5 Decisions on the total value of a technical system are mainly made in the design phase.

11 Developments in Technical Systems in the Course of Time *(Phylogenics, Evolution of Technical Systems)*

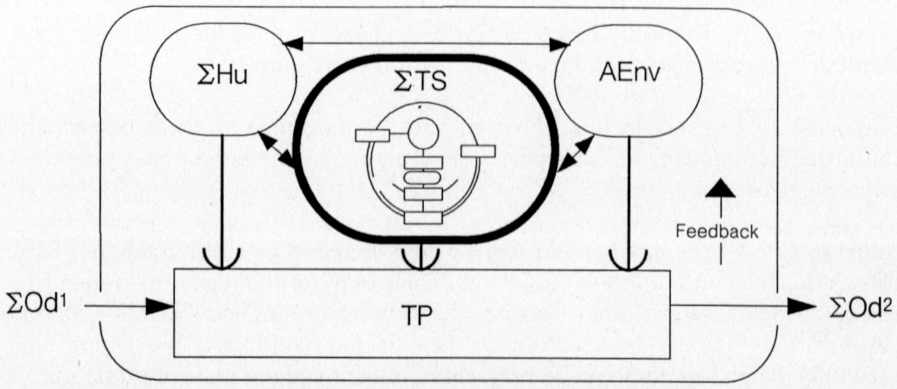

11.1 Regularities in the Long-term Development Process

11.1.1 Developments of Technical Level in the Course of Time

If we consider several technical systems that are intended to fulfill similar functions, and that were produced at certain successive times, we can readily observe the many changes that have occurred. A well-known example is the motor car. Its historic development is illustrated in figure 11.1 (compare [107]). A much smaller variation of form is displayed by large Diesel motors, see figure 11.2. Each of these technical systems has essentially remained true to its main purpose, a car is still a car, and a Diesel motor is still a Diesel motor. Nevertheless, a series of properties of these technical systems have altered, as a consequence of the increase in the perceived needs and advances in knowledge and the means of fulfilling the needs. This is not only true for the appearance properties of these systems which can be readily observed, but also for other less obvious properties such as speed, acceleration, safety, power, weight, etc.

This context is expressed in the term *technical level* applied to a technical system at a certain time. It signifies the sum of the technical properties and their measures and parameters (including performance) which corresponds to the technical value of a system as previously defined. The technical level shows two major aspects, how close to the state of the art the system was at the time of its manufacture, and how good its reliability and operational properties are.

Development thus has two major aspects. One (treated in chapter 10) is concerned with a particular technical system, and its progress from idea to realization and maturity for marketing at a certain time. In this view, development (singular) takes place as a part of the design and manufacturing process, and is intended to raise the product towards an acceptable technical level, and closer to the state of the art existing at that time. The second aspect concerns the developments (plural) over a longer period of time of the technical level and the state of the art for a suc-

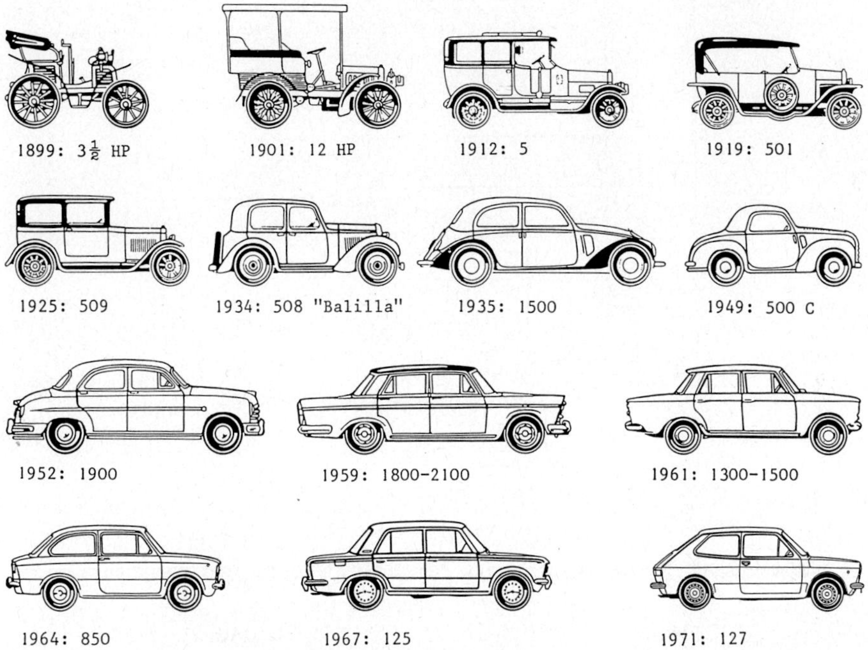

Fig. 11.1 Developments in Time of a Series of Passenger Cars, FIAT Co. Italy, 1899-1971

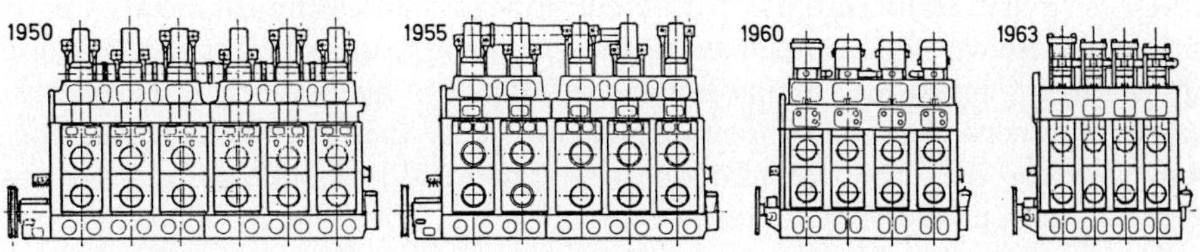

Year of Construction		1950	1955	1960	1963
Power Rating	HP kW	6800 5060	7000 5220	6640 4940	7200 5360
Length	ft m	57.5 17.5	52 15.9	33.5 10.2	28 8.5
Mass	tonne	404	395	260	240
Price	£/HP £/kW	43 57.7	32.2 43.2	22.9 30.7	18.7 25.1

Fig. 11.2 Developments in Time of a Diesel Motor, William Doxford Co., England, 1950-1964

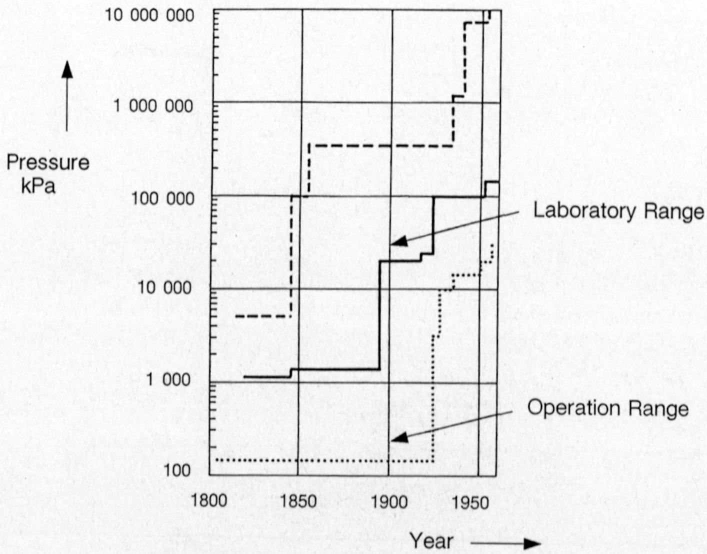

Fig. 11.3 Development of Limiting Measure of Pressure in Laboratory and Operation

cession of technical systems to fulfill a given purpose and function. These developments are the subject of this chapter.

Consideration of individual parameters reveals further aspects of the progress of developments in technical systems over a time period. Figure 11.3 shows the trends for maximum working pressure of liquid media used in energy and process equipment over a period of time, both in the laboratory and in industrial operations. The major difference from the previous examples is that increased pressures have been achieved by employing different principles and technical means, i. e. by different technical systems that have as their common feature the abstract function of "containing pressure". Similar trends can be seen in the developments of transportation speed. Figure 11.4 demonstrates not only the general speed relationships, but also the development of speed in individual types of transport systems (e.g. cars, propeller-driven aircraft, jet aircraft, etc.). According to figure 7.2 these various families of technical systems can be classified under the common abstract function of "transporting humans". Each individual family shows a development curve for speed that has a characteristic shape. They show an asymptotic approach to an upper limiting value which is expressed by a law of nature, for instance the maximum attainable aircraft speed from thrust generated by aircraft propellers, or limits imposed by properties of materials such as strength. A special case of these limits is conditioned by the environment of the technical system, for example the speed limit that is caused by poor road surface. In such a case, the development curve need not approach the limit asymptotically, it may even intersect the limit, and the limit may also be subject to a trend in time.

The serial members of developments of a technical system are often divided into "generations", signifying some major observable change. Some changes from one generation to the next may be seen, for instance, in figure 11.1 (the change

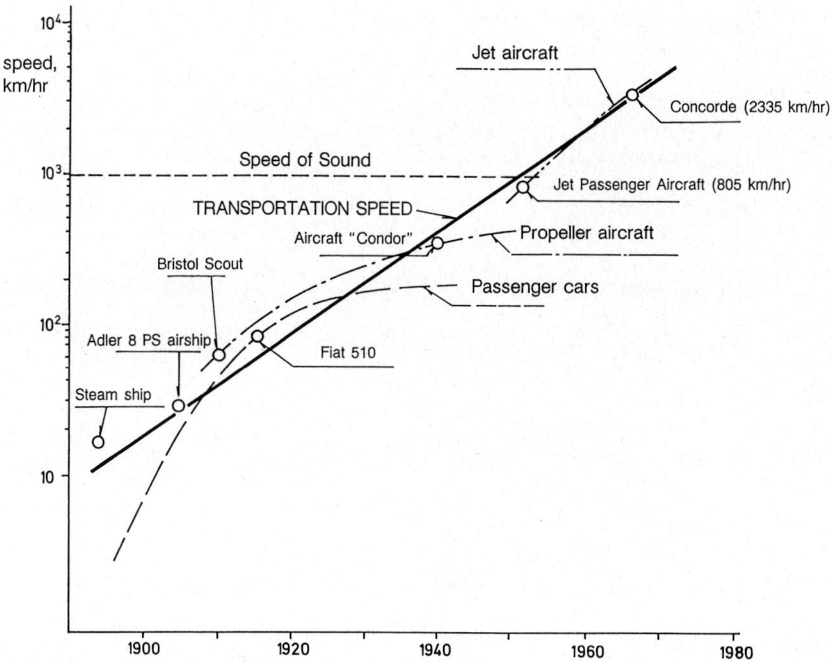

Fig. 11.4 Developments of Transportation Speed

from a separate frame up to the 1952 model, to a load-carrying body shell from the 1959 versions onwards) and figure 11.2.

Figure 11.3 also shows that at any one time various measures of the parameters exist, e.g. different measures of working pressure are commonly achieved in laboratories and in industry. In general we may expect further measures to appear which are related to development and manufacture of products. Investigating the technical level, including the measures of properties and parameters, of a certain family of systems, we obtain various development curves, and indications of the technical level existing at various times and under various conditions, as shown in an idealized form in figure 11.5.

The general growth curve in figure 11.6 indicates that developments over extended time periods in biological systems are very similar to those of technical systems. Similar curves are found to exist for market saturation, cost vs. time for large projects, and other related phenomena.

11.1.2 Developments of Consumption of Technical Systems in the Course of Time.

An important quantity that varies strongly with time is the rate of usage or consumption of technical systems. The usage curves could be transformed into production quantity curves, but for simplicity we will merely assume a similar trend in these quantities. The increase and decline of sales (as a measure of usage or con-

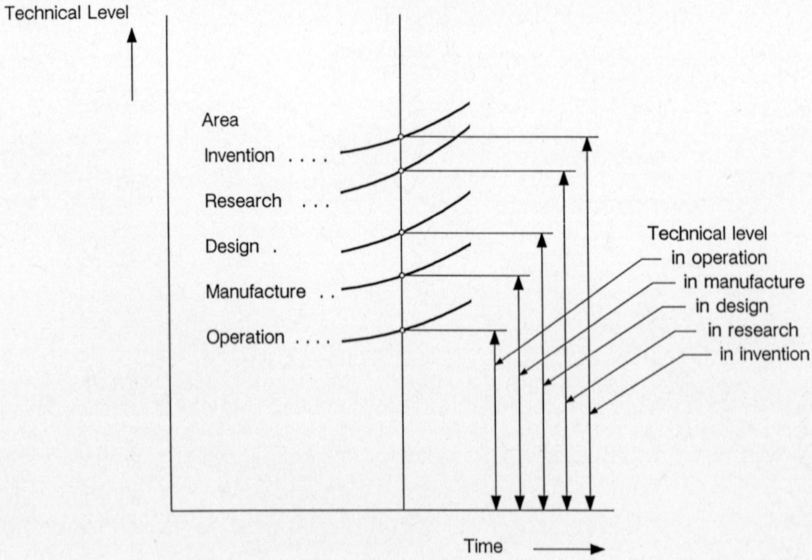

Fig. 11.5 Idealized Development Curves for Some Areas of Activity, and their Technical Level

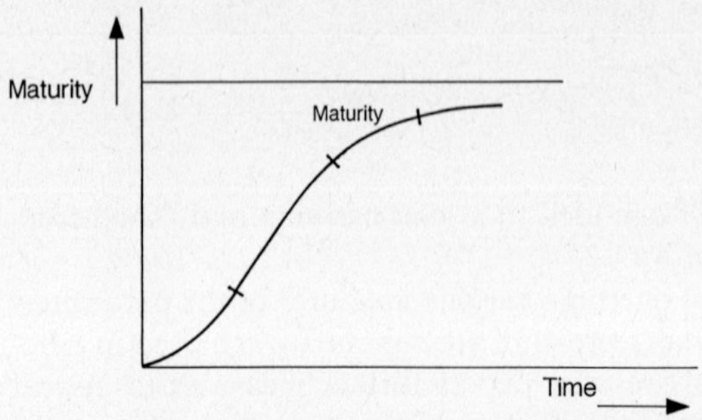

Fig. 11.6 Development Curve of Biological Systems

sumption) of a new product, in our terminology a technical system, can occur as shown in figure 11.7-A. This somewhat irregular curve can be replaced by a simpler curve shown in figure 11.7-B to reveal a general trend. The end of this curve demonstrates that the usage of a type of technical system must not necessarily drop to zero when a new model or product appears. The older system (even if convention or competition declares it "obsolete", see section 7.2.7.3) is still used in reduced quantities, for instance the emergence of combined harvesters has not fully displaced reaper-binders and threshing machines.

Usage curves can also be displayed using other variables on the ordinate axis. Showing the cumulative sum of all products manufactured to date (the integral curve to figure 11.7-B) delivers a growth curve similar to figure 11.6. Alternatively, the difference of consumption over a suitably small time interval can be plotted, typically resulting in the differential curve shown in figure 11.7-C.

These curves represent only the ideal case. Various influences modulate the real curves in different ways. Such influences include the technical solution itself, com-

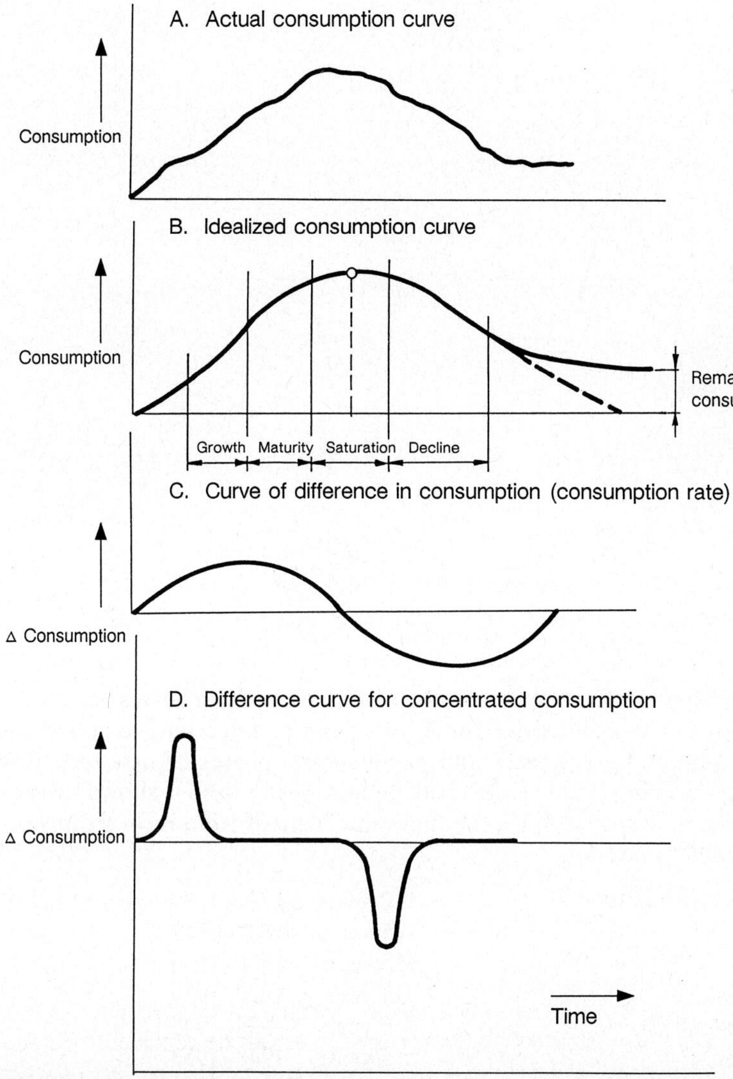

Fig. 11.7 Consumption Curves

peting solutions with identical or related functions, natural influences, or political and social effects (embargo or crisis situations). A period of rapid introduction, followed by stable consumption, and followed in turn by rapid decline, could result in the positive and negative peak in the difference curve of figure 11.7-D, and depends on the influences of market forces. Figure 11.8 present an example of real consumption for three consecutive models of tape-recording machines.

If we investigate a number of usage curves of successive models of a technical system, we will usually find that the wave-length of these curves is not constant, it reduces. The usage life of products (from the manufacturer's viewpoint) is becom-

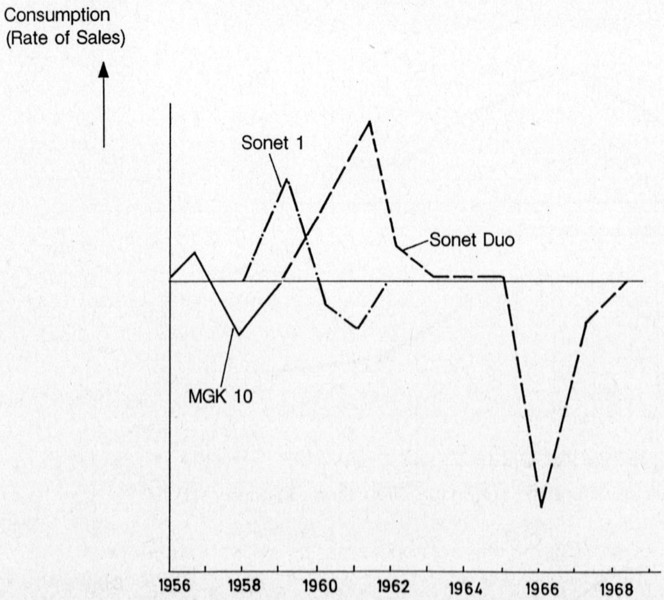

Fig. 11.8 Example of Actual Consumption Curves for Three Successive Types of Tape Recorder

ing progressively shorter. Simultaneously, the amplitude of these curves is increasing as consumption or sales increase. These facts point to a necessity to accelerate the origination (design, development and manufacture) phases of new technical systems. That this necessity (on average) may be historically valid is shown by typical data for some representative TS, the times that elapsed from recorded invention to a marketable product:

photographic cameras	112 years	1727–1839
radio	35 years	1867–1902
radar	15 years	1925–1940
television	12 years	1922–1934
atomic bomb	6 years	1939–1945
transistor	5 years	1948–1953

Some products appear to reveal an opposite trend, note the reversal of dates from television to radar. There is also some doubt whether comparisons of these particular systems are useful, and what events constitute the start and the end of that time period.

11.1.3 Factors of the Process of Developments of Products in Time

The development in time of a family of technical systems, as well as the more obvious factors of new manufacturing methods, new materials and new markets, characterize the process of innovation.

Developments over longer times serve to perfect certain properties of existing products from one model (and manufacturing series or even batch) to the next, and to introduce new products, with the aim of satisfying the increasing needs of society. These needs change with developments in and of society. In the past, the declared needs for products reflected almost exclusively the problems connected with survival and maintenance of life. Current needs frequently go far beyond these essentials, compare chapter 2.

In addition, the working conditions and the environment which existed when the developments in these technical systems were started are subject to change. New insights in knowledge available from experience and science are gained by developments, and these enable the use of new materials and improved methods. These *technical capabilities* can only be fully utilized if the economic situation permits the developments to be financed, and those improved technical systems to be made. It also requires some *motivation* for society to urge solutions to the problems. This motivation can emerge either from categorical demands of society (e.g. satisfaction of the needs for national defence), or from demands of the economic sphere (e.g. the profit motive), or from developments in experiential or scientific knowledge (e.g. attempts to implement new insights, to alleviate the effects of pollution, etc.).

When investigating how one of these developments depends on another, we must not forget that the developments in time of society, science and economy are interdependent. This system of relationships is plotted in figure 11.9.

If these relationships are to be listed in a more complete way, the *internal factors* of the design processes that realize developments in products must be included.

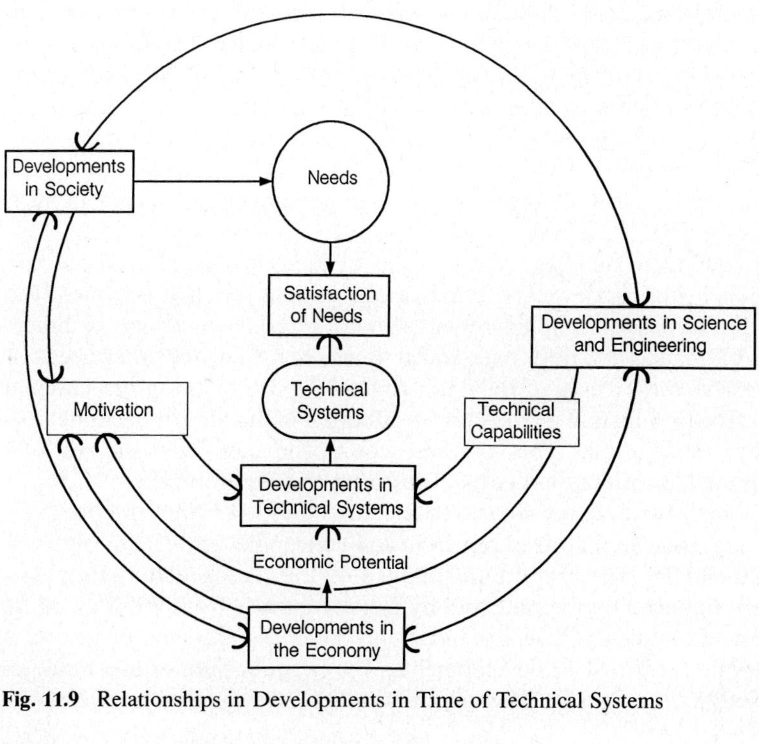

Fig. 11.9 Relationships in Developments in Time of Technical Systems

The following factors are active (and will be discussed in more detail in chapter 12):

(a) designers,
(b) technical information (in a relationship with "technical capability"),
(c) technical working means,
(d) management and
(e) environment conditions of the design process.

A further factor that influences developments in successive technical systems and the satisfaction of needs is the situation concerning availability of raw materials. Are oil, coal, uranium, iron, nickel, titanium, etc. available from the country's own resources? Even where these questions are not of overriding importance in peacetime and under conditions of good economic cooperation, such considerations may be justified.

The risks attached to developments of products also play an important role. For instance, statistical investigations have shown that in the U.S.A. of 100 conceptually solved problems only 10 solutions were realized, and only three of these resulted in new products that maintained their place in the market and were suitable for further developments.

11.1.4 Execution and Scope of Developments and Research

As requirements on technical systems become more severe, together with growth in the numbers of technical systems, and reduction in average product life, the scope and extent of basic and applied research, and of conceptual and technical developments of technical processes and systems, must increase proportionately. A characteristic number usually used to measure the extent of this work is the *number of experts* (scientists, engineers, technicians, etc.) employed in these tasks. Currently, this number is estimated as 4 million world-wide, approximately 0.1% of the world population. In the industrialized countries this proportion is much larger, statistics show that 6.2% of the U.S.A. population is employed in such work.

Another characteristic number for the extent of these classes of work is cost. The industrialized countries expend an increasing amount for such activities. The expenditures for basic and applied research, and for conceptual and technical developments of TP and TS, have risen much more than those for other needs of society. At present these costs account for about 2 to 3% of the gross national income. An interesting feature of the current situation is the almost accurate proportionality between various types of these costs. The cost for basic research amounts to about 10% of the total costs of research and development.

How can these costs be covered? Increasing costs and risks limit the possibilities of performing basic and applied research, and conceptual and technical developments of TP and TS. The proportion supplied by the state generally increases. Universities are financed by the state and by direct contracts from industry. More universities are engaging in basic research, with some departments or institutes involved in applied research or developments. The negative side of this mode of operation is seen in the usually perceived methods of selecting research themes, guided more according to personal interests than the needs of society. Availability of funding and the criteria according to which such funds are allocated "drives" the research effort, especially in engineering (but much less in the pure sciences, mathematics and the arts and humanities).

A further indirect support for basic and applied research, and for conceptual and technical developments, is provided by the country or region and its government by its financing of organizations for information and advice (libraries, facilities for cataloging and searching abstracts, research associations, specialist information centers, etc.). These sources of information are made available to interested persons and organizations against payment of formal fees or an effective net cost.

The high costs, coupled with a lack of suitably qualified personnel, enforce a reasonable increase in effectiveness of the efforts towards basic and applied research, and conceptual and technical developments. Cooperation on as broad a basis as possible is desired, at national and international levels, because practically no country by itself is in a position to successfully conduct the necessary research or developments in all fields. Solving such problems currently demands a high level of specialization (but also good general knowledge). A noticeable trend is found in the increase in the number of international scientific and research organizations, and particularly in exploiting foreign knowledge through acquiring licences and "know-how".

11.2 Trends in Developments

In order to establish directions for developments over longer time periods in various fields of engineering we must start from the goals of society. This general goal-setting is not easy, it must take into account the contrary interests of various nations, societal classes, occupational groups, political streams, individual people, and so on.

The fields of philosophy and sociology, which are concerned with formulating societal goals, appear too foreign and abstract to engineers. This reaction of engineers is incorrect and inappropriate, the engineers' responsibility for developments of and in society is becoming much more similar to the medical profession's responsibility for human health. The code of ethics of professional engineers is one of the means that can contribute to avoiding mis-use of science and engineering to further the interests of individuals. The circumstances addressed by the codes of ethics are complicated by the fact that many engineers are employed by companies and corporations, and are therefore not totally free to follow their conscience in applying ethical considerations. Even the self-employed engineers and consultants must take market pressures into account. The economic and market systems are thus important factors in operating the socio-technical system, and affect among many others the persons (and their scope for action) who research to obtain information, and who design and develop TP and TS.

Several societal goals are undisputed as ideals for the desires of humans in general. Without implying any priority, probably the most important of these are:

eliminating starvation
eliminating war
eliminating diseases and prolonging human life
eliminating crime
improving the standard of living
increasing the level of education
reducing physical labor
shortening the time spent in work.

How far these goals can contribute to the satisfaction and happiness of humans is a different question. Opinions differ about the proportion of participation that engineering and science should have in attempting to realize these goals. The views range from a complete rejection of engineering and science to encouraging their most rapid development. This reflects rather the quantitative side of the problem. It is qualitatively important to recognize what influence engineering and science, and the means they have at their disposal, can have on the happiness of humanity. In order to effectively and correctly guide these developments we must decide in what direction they should progress, and how they can be regulated.

Many readers will probably think that these considerations have already gone too far in the context of this book. Nevertheless, all engineers should be aware of the consequences of their work, and should always challenge their own tasks by comparison with the disputed interests of humanity.

Trends in long-term developments of TP and TS are frequently formulated on the basis of particular properties of technical systems. Accepted directions of development include the following major trends:

- **mechanization** - involves progressively transferring working functions from humans to technical systems;
- **automation** - involves progressively transferring regulating and controlling functions from humans to technical systems;
- **computerization** - includes transferring routine decision-making from humans to technical systems, usually adding some functions of monitoring system performance, and progressing towards expert systems and artificial intelligence;
- **electrification, chemical and bio-technological** tendencies imply enhanced use of these other phenomena as working principles for solving technical problems;
- introduction of **advanced processing and manufacturing methods** based on newer working principles (chemical, biological, managerial and others).

If we compare these trends with the societal goals discussed above, it is clear that their only direct cffect is that of reducing the physical work of humans. The connection to other goals is felt indirectly.

A better overview of the means that serve in attaining specific goals can be obtained by characterizing the developments over time in important technical fields. The current spearhead of technical endeavors of the major world powers resides in:

- information technology
- space technology (rocketry, etc.)
- atomic energy

- penetrating the ocean depths
- environmental protection
- medical technology
- development of transportation and distribution facilities
- personal transportation
- interplanetary exploration

Generally these activities attempt to ensure better contacts between humans, to unlock new areas for humanity, and to enhance human health.

Total developments in a family of TP or TS are determined by developments in each and all the properties of technical systems, but in particular by developing the design properties - the external properties are established by decisions about the design properties, see chapter 7. The directions of developments of individual properties are designated by certain terms as outlined below.

It is obviously difficult to give general characteristics for developments in all technical systems. By outlining the trends for a certain class of technical systems, or a concrete system, these trends can be demonstrated as examples. The following descriptions cover all the classes of properties as defined in chapter 7, especially figure 7.1, and show specific examples of development trends from the recent past for each property:

Function:

- **Working functions** are rapidly developing, reflected in developments in processing and manufacturing technologies based on scientific principles.
- **Propelling functions** seem not to have been affected by any great changes. The electrical, pneumatic and hydraulic systems that fulfill these functions are being continuously improved in both technical and economic aspects.
- The **regulating and controlling functions** have seen intensive developments, with a continuing trend towards automation and computerization among the main directions of these developments. Based on electronics and cybernetics, the means to fulfill these functions have reached a high level and are being continuously improved. Important items of improvement range from semi-conductor technologies for a.c. speed controls for motors, to artificial intelligence and expert systems for decision support.
- **Functionally Determined Properties**. The quantifiable properties in this group are characteristic for the value of a system. Pressure, temperature, power, speed and other performance measures have progressively increased in almost all products. The unit size has also generally increased to obtain better economy and efficiency, which may be especially noticed for turbines, freight ships and aircraft. Some noteworthy decreases in unit size have recently emerged.
- **Operational Properties**. Large changes are visible in this group, but not always towards increases. If for instance the demand for reliability increases, a trade-off to shorter working life is usually necessary. Other trends are towards miniaturization of elementary technical systems, reducing dimensions and weights. Operation and maintenance are aided by trends towards modular construction and encapsulation.
- **Ergonomic Properties**. This category belongs to the properties that have been increasingly demanded in more recently times. A strict distinction must be made

between the justified requirements to protect human health (physical and mental), and an exaggerated quest for originality. It is to be expected that scientific investigations of ergonomics will continue to reveal new requirements for TS, by providing information regarding capabilities of humans, and effects on their health.

- **Appearance Properties**. The importance of this category has increased. Only in a relatively small group of technical systems can these properties be ignored or neglected. Aesthetic aspects of products must be given adequate consideration, at least for reasons of competitiveness. These constitute relative and subjective viewpoints that undergo many changes.
- **Distribution Properties**. More efficient utilization of the volume of transportation spaces (ship holds, rail carriages, aircraft holds, packing crates, etc.) is demanded because costs of transportation can be very large in view of the current state of concentration of manufacturing. Similarly the requirements for packaging are increasing, not only from the viewpoint of transporting and climate (e.g. packaging for the tropics), but also for purposes of advertising – the consumer should be impressed by the packaging. Rapid installation and commissioning remains one of the important requirements.
- **Delivery Properties**. A typical slogan for today is that rapid delivery is essential, and this is likely to continue into the future.
- **Law and Standards Conformance**. The increasing international relationships between economic regions make the task of satisfying all legal regulations much more complicated. This concerns especially laws, rules and standards with international validity, and the situation with respect to patents, both of which must receive continual review. The responsibility of the manufacturer for the quality of products is becoming wider. Especially the legal implications of product liability rulings by courts of law can have noticeable effects on producers.
- **Manufacturing Properties**. Increasing use of automatic handling (robotics) and inspection, as well as computer control of machine tools, forces changes in the form of components to enable easier holding and machining. Technical systems should be increasingly powerful, and have better performance, which makes greater demands on production properties of TS.
- **Design Properties**. Successful developments in the internal properties are the condition for realizing all other trends in developments. The internal properties (and consequently the external properties) are determined by means of the design properties (compare figure 7.15), and therefore developments in the design properties must receive maximum attention.
- **Structure**. Developing the various structures of TS depends on developing their elements (process operations, functions, organs or components). At the component level, it is an undisputed fact that the assortment of available machine elements is continuing to expand, and their quality is increasing. An important trend in designing constructional elements is aggregation of functions (otherwise known as accumulation or integration) with the aim that one constructional element should fulfill several functions. Various trends are visible in spatial arrangements and relationships of technical systems. Prime among them is the effort to form modular constructions, arrangements in which many different variants of technical systems can be assembled from prefabricated (sub-assem-

bly) groups. Such groups can then be manufactured in larger quantities. A further direction leads to encapsulation (into cartridges and cassettes) to simplify manipulating replaceable elements of a piece of equipment. Typical examples are audio and video tape cassettes, and film cartridges.

- **Form**. The main trend is to simplify and unify (standardize) forms because form significantly influences the production costs. Such considerations are also important, for instance, for the form of information recorded on computer media. For hardware, this development is somewhat problematic from the aesthetic point of view . Other factors have also to be considered to approach optimization which influences form.
- **Dimensions**. In the group of operational properties above we referred to this variable and the trend towards minimum dimensions (sizes) that should be considered where other requirements do not restrict the choices. The principles of standardized dimensions (preferred size series) and typification of technical systems should be respected (see also section 6.6 and figure 6.7).
- **Materials**. Developments in materials offer the designer increasing possibilities. Four major trends of development in the materials area are observable:
  (1) Expansion of available types of metallic materials. Only 7 metals were in general use up to the 17th century, this number increased to 17 at the end of the 19th century, and today more than 50 metals are generally used.
  (2) Use of artificially produced materials. These can be either alloys (some thousands are available) or plastics and other synthetic materials.
  (3) Improvement of material properties. The ultimate strength of available high-strength steels has increased from 300 MPa to 700–800 MPa. Newer alloy formulations have shown improvements in other properties such as ductility, for example the HSLA (high strength low alloy) steels.
  (4) Use of composites. Many components are built up as sandwiches made of layers of different materials or composition held together by adhesives, such as plywood, honeycomb sandwich plates, etc. Others are made as lay-ups, or as injection or transfer mouldings of synthetic resins, with random fibers, rovings or woven fabrics as reinforcements. Many components (sub-assemblies) are built by using adhesive bonds.
- **Surface Quality**. The quality of surfaces improves with new possibilities of manufacture. Most of these processes also result in lower production costs.
- **Tolerances**. Possibilities exist of achieving closer dimensional and geometric tolerances more economically with the increasingly powerful (and consequently more rigid) machine tools currently available. Computer control of such machine tools may also help to reduce the variability of tool positioning, and enhance the tolerance capability of the process. The general rule of allocating the widest possible tolerances is still valid.
- **Economic Properties**. In order to stay competitive, the manufacturer must keep self-costs to a minimum. The means to achieve good economics are rationalization, standardization, typification, re-use of existing systems, etc. A recent trend has been the introduction of the concept of computer-integrated manufacturing CIM [198].
- **Liquidation Properties**. Some large-scale accidents (e.g. Amoco Cadiz, Torrey Canyon, Three Mile Island, Chernobyl, etc.), and hazards in some waste prod-

ucts (e.g. from long-life nuclear radiation, damaging chemicals) have resulted in new thoughts on disposal of technical systems. Potential shortages of raw materials, and concerns for the environment have increased the focus on re-cycling.

The most important directions for long-term developments in families of technical systems, as mentioned in the context of the classes of properties, may be summarized as follows:

- Economic construction: approach to minimum manufacturing costs.
- Minimalization: trend to achieve the smallest dimensions (e.g. minimum size) of the system for a given function.
- Light-weight construction: trend to achieve the smallest or optimal weight of the system for a given function.
- Composite construction: composites of various materials with advantageous properties.
- Aggregation: trend towards achieving many functions with one function carrier or component.
- Mechanization: allocating the propelling functions to the technical system.
- Automation: allocating the regulating and controlling functions to the technical system.
- Computerization: allocating routine deciding functions and performance monitoring tasks to the technical system.
- Encapsulation: collecting replaceable elements into a more easily handled cassette or cartridge.
- Modular construction: connecting the constructional elements into suitable groups from which many variants of technical systems can be assembled.
- Standardization: unifying the dimensions, forms and other properties and values of technical systems.
- Typification: establishing the optimum number of variants and sizes of selected characteristic properties of a particular kind of technical system.
- Re-use of existing technical systems: using systems that have already been designed and tested for a previous application as some of the component parts of the new system.

11.3 Regulating and Controlling the Process of Long-term Developments

From the discussions presented in the previous sections one can characterize the developments over time in a family of technical systems as a lengthy and expensive process. The complicated relationships that act in this process are shown in a simplified form in figure 11.9. With technical systems of higher degree of complexity, the long-term development of the total system is also dependent on trends and developments of the sub-systems. The results of the process of long-term technical development are also influenced to a large extent by the technical and economic situation of the enterprise.

Under these circumstances it is unthinkable that the process of long-term developments can proceed spontaneously, particularly because we aim to perform it as effectively and quickly as possible.

As with all other processes, the process of development in time receives effects from the regulating and controlling functions of the technical system that drives it. The technical system in this instance includes the administrative and economic functions of society and/or region, as outlined in the discussions of socio-technical transformation systems in section 3.3. This implies that the work performed in locations dedicated to basic and applied research, and conceptual and technical development must be influenced by various direct and indirect (administrative and economic) instruments with the aim of achieving the best results and greatest effectiveness of the processes of development and technology transfer. In principle, there are two ranges of problems, namely one of setting national, regional and local (e.g. company) goals for the necessary developments, and for research, and one of fulfilling these goals. The nature and scope of the tasks of *regulating and controlling* developments, and the accompanying research, depends on the level in the socio-technical hierarchy at which it acts.

First consider the types of regulation and control at national or state level. In spite of the peculiarities of individual countries and their form of governmental organization, some form of regulation or direction takes place in most countries, and at lower levels in states or provinces, regions, counties, cities, etc. Within their own competence, the government organs clarify the needs and possibilities for long-term developments. They then formulate general guidelines (or policies) according to political decisions for the directions of such developments, and at times even define important concrete tasks. Two broad types of government organs can be distinguished, those with executive competence, and those with advisory duties (e.g. research councils). Frequently an independent Ministry of Science and Technology (or its equivalent) exists in addition to these organs. Such organs have a strong position in countries with a planned economy. At the other extreme, a role for planning is almost denied in some countries, and the policy directing tasks are permitted to be undertaken by lobby groups who pressure for their own interests. The time constant for dynamic response, such as changes of policy or direction in these governmental systems, tends to be very long, which may account in part for the cyclic dynamic instabilities in the economies of some countries, and the world-wide effects on trade and commerce experienced in recent decades.

In an enterprise, whether a large multi-national company or a small production or service industry, regulating and controlling the trend of technical developments is one of the important tasks. Formulating the "technical politics" of the enterprise, including/setting corporate goals and general policies, is a deciding factor for its success.

As mentioned earlier, there are two ranges of regulating and controlling actions, namely (a) setting goals and formulating tasks, and (b) monitoring (controlling) the execution of policies towards realizing and achieving the goals. Goal setting and task formulation must be supported by long-term forecasts. The policy-making bodies must have an accurate picture of the desired situation in 5, 10 or more years at the time of setting the goals, and must keep these goals under continuous review. The problems of regulating and controlling the socio-technical systems are difficult to summarize in a short space, the discussions are therefore limited to a few references concerning techniques of regulation and control:

- Decisions should be built on objective facts that have been obtained from statistical and scientific investigations. No forecast is permitted to be based only on intuition. It is agreed that collecting the necessary information is not easy, particularly because one must work for the unpredictable future on the basis of prior knowledge and personal philosophies. It is thus all the more important fully to utilize the available instruments, including statistics and probability calculations using digital computers. The extensive storage capacity and speed of modern computers enable processing of large amounts of information. Several variants of proposed solutions can be prepared, which can make a decision more objective and transparent, *providing* that the algorithmic models are *sufficiently good.*
- The problems of long-term developments in technical systems must be treated holistically. The technical system should not be considered in isolation, but the whole social and technical transformation system including its active environment must be investigated. The relationships summarized in figure 11.9 can be used as basis, and all factors illustrated in figure 10.1 should be taken into consideration.
- In technical developments, cooperation between different organizations and regions is also necessary. One should not, even on a regional basis, try to design and develop everything for one's self. By using the expertise of organizations specializing in basic or applied research, or in conceptual or technical developments, the effectiveness of the long-term development process increases. Frequently the problem can be solved better and quicker by transferring the developments to a specialized development institution, or by acquiring a licence. This latter procedure also has its dangers and draw-backs, it may lead to a loss of local expertise.

Regulation of long-term technical developments is performed and executed with the help of *plans* which establish trends of development, and specify concrete tasks and deadlines for individual countries, regions, organizations, companies and departments. If these plans are correctly formulated, they guarantee coordination of the work. Control is performed by daily monitoring of work, and if necessary changing the planned tasks.

11.4 Motivation for Long-term Developments

Possessing technical capabilities and having economic possibilities are not sufficient conditions to execute a program of long-term technical developments, there must also be a motive that acts as driving force.

If there is no interest in the solution of a technical problem it will not be solved. The sum of these technical, economic and motivational components must in general attain a certain minimum magnitude to set the train of developments in motion. In extreme cases this implies that strong motivation can influence the technical and economic situation, and technical or economic potential can create a good motive for action.

Crisis situations in society provide a particularly strong motivation. Looking back on the history of technology shows that great technical advances have been achieved particularly during wars. The American civil war of 1861–65 ("the war between the states") was the first occasion of use of "ironclad" warships, and a submarine. The first world war 1914–1918 saw the introduction of the armored tank, and stimulated major developments in internal combustion motors. The second world war 1939–1945 brought further important results of technical development such as gas turbines, rocketry, digital and analog computers, remote sensing (e.g. radar), and exploitation of atomic energy.

The causes for the increases in technical developments during wartime are understandable. In one sense it is the desire for victory, but also the distress situation of peoples has forced them to restrict their consumption and re-direct their means mainly towards developments in appropriate families of technical systems.

Apart from wars, long-term developments have been accelerated by catastrophes and disasters. The large numbers of deaths and disablements resulting from epidemics, large accidents involving technical systems for road, rail, sea and air transportation, fires and gas explosions, nuclear plant, mining disasters, and natural disasters such as earthquakes, volcanic eruptions, hurricanes, etc. served as impulses to improve existing technical systems, or to create new ones.

If we observe the "peaceful" competition of the major powers, whether in space exploration or in other technological fields, we must hope that humanity has become wiser by its collective experience and no longer need the impulse from human-initiated catastrophes for further technical developments.

In spite of the wish expressed in the previous paragraph, statistics show that cumulatively large number of deaths occurring in widely distributed small accidents tend to be disregarded, especially compared to similar or smaller numbers of deaths in concentrated happenings. Some physical principles are regarded as "more dangerous" than others, even if the available evidence does not support these impressions. It is also clear that some of the relationships between causes and consequences are indirect, or only come to light after very long periods of time. Examples range from the insecticide DDT, toxic dioxins, asbestos, and chlorine compounds, to nuclear radiation.

People are usually aware that they take risks throughout their lives. If the risks are imposed on people, through employment and other factors, they are only accepted if sufficiently small. The same people will voluntarily submit themselves to much larger risks in the interests of entertainment and pleasure – most sports and many recreation activities fall in this category, and equipment needed for these activities provides an intense motivator for developments in technical systems.

It is also true that at times an invention will produce a potential technical advance for a problem that has not been fully or formally recognized. This circumstance leads to "a solution searching for a problem". For instance, the transistor was invented at least on five previous occasions, at intervals of a few years, as demonstrated by publications and patents. It was only on the last occasion that the "solution" of the transistor came at the right time to meet its "problem" and then its technological and economic development could proceed. Similar historic facts have emerged for many inventions – vehicles driven by high-speed internal combustion motors that precede the famous Daimler car are found in museums in various countries (e.g. Siegfried Marcus in Vienna, Austria, his first motorized cart ran in 1864/65, a second version in 1875 [95]); sewing machines that precede the famous Singer patent exist (e.g. Joseph Madersperger, also Austrian, about 1807 first prototype, 1815 second version [95]). In fact, multiple independent invention of a technical process or system for a certain function is no rare event. Almost every new concept of technical system for which "the time was ripe" was invented in several places about at the same time. It is in human nature that the inventors of one's own country are usually only acknowledged after their death, and the names of inventors from other countries are either suppressed or unknown (e.g. in most English-language countries, the term used for the common SI temperature scale is degrees "centigrade"; it was defined by Anders Celsius, a Swedish scientist, and non-English people, and SI, have acknowledged it as the degree Celsius).

In most such cases, the developments in families of TP or TS were inhibited by political decisions, technical ignorance of the governing groups, or lack of recognition of the "problem" for which the "solution" was useful. Many TP or TS exist that would help in a different situation, but it usually requires a "product champion" to accept, try out, and report on successful usage, before the existing TP or TS can be accepted for the new task. Thus the history of technological developments, coordinated with the developments in societal structures, morality, politics, etc., is a rich source of information, and adds to the engineer's humility and diffidence.

11.5 TS Developments in Time – Statements

Statement 11.1 Technical long-term developments should support and ensure humanitarian goals of society. All engineers, technologists and technicians should do as much as is in their power to ensure that technical developments serve these societal goals, and are not misused in the interests of individuals or parasitic groups.

Statement 11.2 Developments in technical systems in the course of time consists of a chain of elementary changes in the properties of technical systems. The cumulative consequences of many small steps can appear as a quantum leap – a technical break-through.

Statement 11.3 The speed of long-term developments of TP and TS is increasing, whereby:

- shorter time from idea to realized TS is demanded;
- shorter effective lives of technical systems become acceptable;
- requirements for certain properties of technical systems change their trends.

Statement 11.4 The long-term developments of technical systems are influenced by three main factors:

- technical capability
- economic potential,
- motivation towards such developments.

Other important effects can come from the raw material situation of a country, and the operators of the design process.

12 Applications of the Theory of Technical Systems

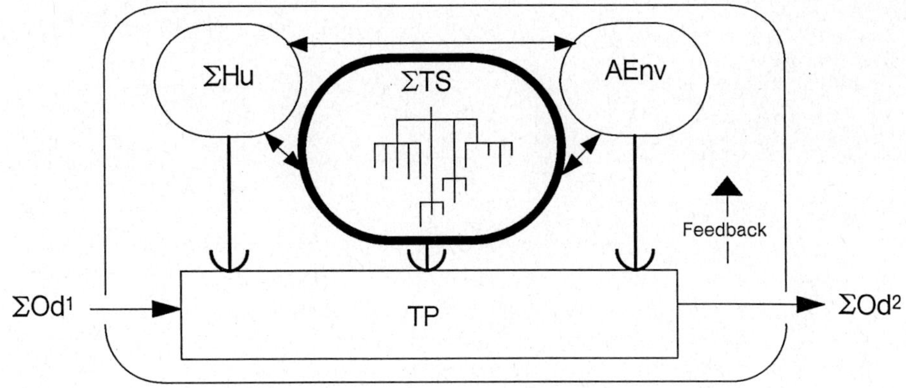

Some questions about applications for the theory of technical systems were posed in chapter 1. Various areas for which this theory can be applied have crystallized from the discussions in this book, and from other efforts related to the theory. They can be characterized by key phrases such as:

- a basis for coordinated and extended insights into technical systems;
- a starting point and coordinating concept for special theories of technical systems;
- a source of knowledge for engineering design and design science;
- a basis for electronic applications of data processing: algorithms, data banks, expert systems and artificial intelligence;
- a basis for an integrating theory useful for educational systems;
- a basis for communication and understanding between technical and non-technical persons;
- a basis for the history of engineering, technology and science.

The first of these applications requires few further comments, this book contains many facts and insights related to technical systems, and some advice about designing. Some comments about other applications appear in the following sections.

12.1 Special Theories of Technical Systems

The initial discussions about types of theories (refer to section 1.5.3) mentioned two categories from the viewpoint of validity: (a) the general theory of technical systems, and (b) various special theories, each dealing with a type or aspect of a technical system. The general theory has been the main subject treated in this volume, with a few excursions into special theories, for instance in section 4.11.

Special theories are concerned with individual categories of technical systems

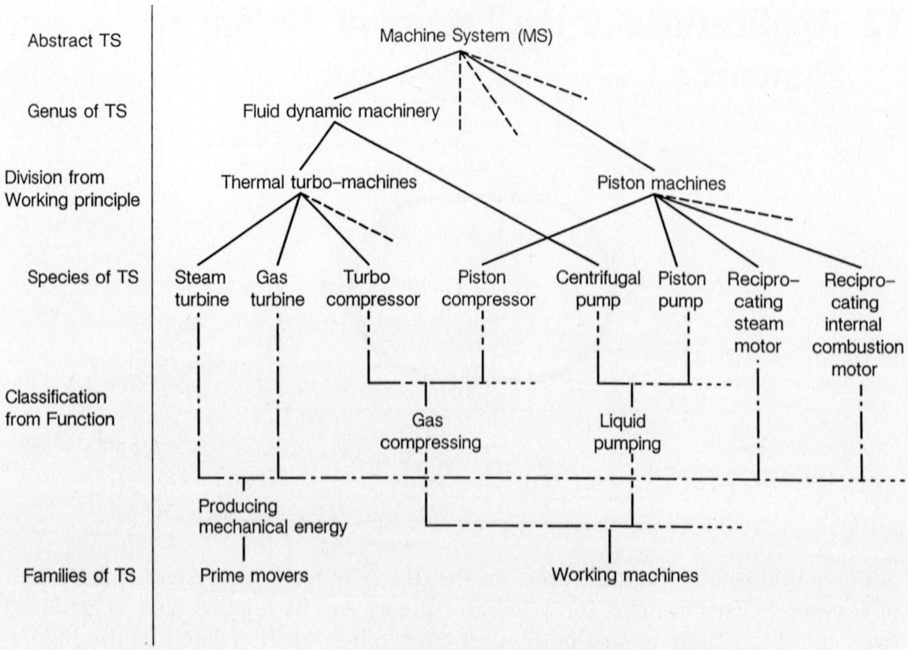

Fig. 12.1 Formation of TS Types - Examples from Mechanical Engineering

and processes. The major possibilities for classifying technical systems from various viewpoints were discussed in chapter 6. Some of these classifications can provide a suitable basis for formulating special theories of technical systems.

Individual categories of special theories built from different viewpoints exhibit overlaps, as demonstrated by figure 12.1 from the field of mechanical engineering systems. It would be possible to find a "theory of steam turbo-machinery" either within a "theory of thermal machines" or within a "theory of turbines". There will, of course, be cross-currents and cross fertilization in many of these special theories of technical systems (a) from various types of machinery, and (b) from established scientific theories such as thermodynamics, fluid mechanics, physics, etc. In fact, the established scientific theories may also be regarded as special theories of technical systems, because they deal with certain aspects of technical systems at a higher level of abstraction. The areas over which these theories range may be wider or, in some cases, extremely narrow.

In analogy with figure 6.8, one of the available starting point for formulating a hierarchy of special theories can be found in the levels of abstraction. An example would be a theory of machine tools (at the "phylum" level), a theory of metal-cutting lathes (at the "class" level), a theory of universal lathes (at the "family" level), etc.

Other formulations of special theories may be developed that deal with lower levels of complexity of TS. They may include theories of mechanisms, linkages, machine elements, and so on, as special theories of components and sub-assemblies of machine systems. Other areas of technical systems would lead to similar examples.

From these examples it is possible to derive some significant criteria about TS from which special theories may be classified, particularly:

- their function (pumping, metal cutting, etc.)
- their mode of action
  (for the function "pumping" - piston, centrifugal, axial-bladed, gear-action, screw-compression, etc.)
  (for the function "metal cutting" - rotating or reciprocating tool, rotating or reciprocating workpiece, geometrically defined or undefined cutting edge, etc.)
- their degree of abstraction
  (for the function "pumping" - pumping machinery, gas compressor, positive-displacement blower, Roots type Diesel supercharger, etc.)
  (for the function "metal cutting" - machine tool, lathe, turret lathe, etc.)
- their degree of complexity (machine, mechanism, machine element, component, etc.)

The purpose of each of these special theories of technical systems is to collect the available knowledge about that special type of TS, and to categorize it in suitable ways. If the questioning and categories are derived from the general theory of technical systems, as laid out in this book, then a more uniform and usable structure of all technical knowledge can be achieved. The advantages are obvious:

- adapting the existing scientific theories into the framework of the general theory of technical systems should expand and consolidate the available knowledge,
- cross-fertilization between different branches of engineering should be much easier,
- experiences from one area should be more easily transferrable to another (both between individual areas, and between practice and research - in both directions), and
- some major failures attributed to "lack of appreciation for one technical area of knowledge available in other areas" may be avoided [208].

The special theories of technical systems are therefore faced with the following complex of questions:

- types of transformation systems where a particular type of TS may be employed - analysis of the elements of the transformation system, including operands, technologies, operators, etc.;
- output effects that may be selected as tasks for TS types - their typical function structures, variations and modifications;
- modes of action and working principles usable or available for use for individual functions, as presented for instance in some design catalogs [187]);
- typical function carriers (organs) and their relationships - organ structures, e.g. some of the information available in the appropriate design catalogs [187]);
- typical properties, design characteristics and their states of embodiment - comparisons, evaluations, quality problems, achieving the desired properties;

- problems of generation (ontogenesis) of the TS type, including designing and manufacturing them;
- experiences of development, and possible trends.

A comprehensive theory that addresses all these questions will be substantially different from the various forms of theories available in current technical books. Major parts of the existing theories will be directly applicable, other parts will need adapting or transferring between areas, still others will need to be newly developed.

The whole system of these special theories of technical systems (coordinated by the general theory as presented in this book), and the knowledge and experience collected in them, will also satisfy the computer experts, because data can be made available in suitable form and suitably categorized and cross-referenced.

12.2 Engineering Design

Designing is a particularly fruitful domain for the theory of technical systems. Nevertheless, applying the theory of technical systems to engineering design is not as easy as may appear.

Investigations about engineering design are of fairly recent origin. Progress can be traced from about the middle of this century [37, 38, 44, 58, 77, 90, 93, 99, 100, 109, 132, 133, 143, 146, 150, 156, 167, 168, 169, 181, 182, 194, 205, 210]. More recent trends have included developing some details and insights about engineering design, and methodical procedures for designing [40, 64, 65, 67, 69, 74, 75, 86, 102, 180, 207]. Many current advances in the science about engineering design follow directly from the theory of technical systems presented in this book [28, 112, 119, 126, 127, 188]. Other attempts have been based on cybernetics and systems concepts [35, 39, 98, 135, 136, 137, 138, 139, 202, 212, 213]. An international exchange is now taking place in various conferences and workshops [30, 68, 116, 117, 120, 124].

When designing, human beings (designers) attempt to define a technical process or system, based on some declared or recognized set of needs. The TP and/or TS is usually an artificial transformation process and/or a technical system. The product should solve a problem such that it can be (a) physically produced and implemented, (b) economically and financially justified, (c) politically, morally and socially accepted. This design process includes human beings and their social context (in the widest sense), and consists of partial processes such as planning, predicting a market, financing, designing, evaluating, etc., and anticipating the needs and problems of developing, manufacturing, assembling, assuring quality, testing, marketing, repairing, maintaining, using, disposing, impacting the environment, etc.

Engineering problems are typically "dirty", ill-defined, with many possible solutions which are difficult to evaluate in any rational form until a large amount of innovative and investigative work had been done. Large systems problems can be solved in principle to deliver "clean" problems. These problems in turn represent "dirty" problems for the next more detailed design phase, which frequently has to be performed by a different engineering speciality – and at the tail end of this hierarchy, it is usually the task of mechanical and production engineers to design, manufacture and assemble the components of the solution. A new TP or TS will only be accepted by the engineering community and the market if it shows both technical and economic advantages.

We can therefore define "Engineering Design" with reference to figure 12.2 as follows [67]:

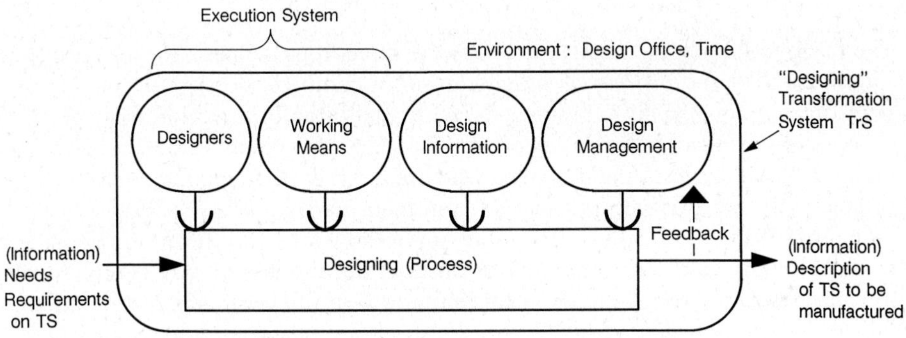

Fig. 12.2 Model of Design Process

Engineering Design is a process performed by humans aided by technical means through which information in the form of REQUIREMENTS is converted into information in the form of descriptions of TECHNICAL SYSTEMS, such that this technical system meets the requirements of mankind.

In other words, "the designer designs a device, product, artifact, transformation, process or procedure in a perceived context". An analogy with grammatical terms shows that the designer is the subject (of the design process), designing is the verb (the activity), the device is the object (to be designed), and the context (organizational, institutional, societal, environmental, goals, etc.) is a conditional clause to the sentence referring to the designer and/or to the use of the device. This definition of engineering design forms a bridge between the theory of technical systems described in this volume, and the theory of design processes (design science) outlined in this section.

A holistic study of engineering design should thus cover a number of factors and aspects, as suggested by the above definition and figure 12.2:

- the object (TP and/or TS) to be designed – the **operand of the design process**, its nature and properties, processes to be performed, necessary influences on the process (from human beings, technical systems and the environment), parts, abstractions and models, life stages, representations, evaluations and decisions, etc., considered as generalizations applicable to all transformations and technical systems, and to particular systems;
- designers – the **human operators**, their characteristics, working methods, use of information;

- working means – the **technical systems**, tools and aids that they do or may use, etc.;
- the activity – the sequence and structure of **operations of the process**, its components and relationships, any systematic or procedural aspects (methods), creativity and intuitive factors, and any other factors affecting design progress, etc.;
- the context – the **information system** and **management and goal system** in which engineering design takes place; and
- the social, moral and political context – the **environment** of designing, producing and using the resulting technical system.

The information about a need (or about the direct requirements for a product) is transformed by the process of designing into a complete description of the proposed means for fulfilling these needs, particularly the information for manufacturing the designed technical systems. The "design system" that performs the design process includes the designers who are actively participating, their working means (tools, etc.), design information, and their management.

The transformation called "designing" preferably takes the form of a sequence of one-to-many mappings from a more abstract form of model of a TP and/or TS to a more concrete form (see for instance figure 5.4). One more abstract model of a TP or TS can be realized and concretized by several proposals and more concrete models of TP or TS. Each such mapping is followed by a series of other design operations. These include evaluating, selecting, improving, quantifying and optimizing the most promising proposals to carry forward into the next (more detailed) phase. They also contain various analytical operations to establish, optimize and verify the appropriate measures of the design properties. These smaller procedures represent the partial transformations and operations recommended for performing the design process (see also [119]). If this procedure is not followed, EITHER a number of the available alternative solutions will not be considered (in favor of jumping to the solution first thought of), OR the number of combinations of alternatives may be subject to exponential growth and combinatorial complexity. There should be as little iteration and backtracking as possible in the design process, but such procedures are generally unavoidable in practical engineering design work.

The transformation of information therefore comprises a series of connected operations with consequent changes in the operand of the design process (i.e. the technical processes and technical systems to be designed) as a mode of conducting the progress in designing. Each step or stage of this transformation may be performed with the help of a suitable **method**. "Methods" have only penetrated into engineering practice to a limited extent, but especially in connection with computer aids to design (CAD) and manufacture (CAM or CIM). The sum of actions in designing, and the sum of recommendations for methods that can be used, is termed **Design Methodology**, a heuristic prescription (and model) of "how to proceed".

Engineering design cannot be performed in isolation. Designers must design SOMETHING, and therefore must use a wide range of knowledge and experience to make that something into a marketable product. Such knowledge is an essential peripheral to the model of the design process (and an operator of that process), and preparing the information for better usage by designers is a vital task. In order to produce good products, engineering designers require a vast amount of knowledge and experience, and must be familiar with the vocabulary (in words, symbols and images):

- of existing processes and products, organized and categorized in useful ways;
- of generalized properties and measurable quantities, phenomena and their behaviors, accuracy, precision and expected tolerances of their measures, etc.;
- of the compositions, elements and their relationships (the constituents of a system);
- of special and social processes - company operation, inter-personal relations, politics, economics, history, communication, process operation, manual and mental skills and abilities, designing, modelling, teaching and learning, etc., including societal values;
- as results of both formal learning and informal observation and experience;
- plus an open mind to allow novel combinations of knowledge, and discovery of new devices and principles.

This provides a basis for directed creative activity. "Creativity" alone is not enough, and it is doubtful whether creativity can be measured in any consistent way.

Two aspects must be clearly distinguished, namely (a) the knowledge that is used by designers during their work - the science and technology FOR design - and (b) the knowledge about the nature of their work - the science OF and ABOUT design. The former has received much attention, and interacts with scientific research and culture. This section discusses some aspects of the science ABOUT design - design science.

The tools that designers use have changed in recent times. Wider application of computers for analytical calculations, graphical representations, and even some heuristic and diagnostic searches and decision-making (for instance by expert systems and artificial intelligence techniques) has extended the range of technical systems that act with and for designers to help them drive their design processes. An improved theory of engineering design should allow larger sections of the design process to be algorithmized and automated, without denying the primacy of human beings in their creative talents. Some progress has been made in limited fields along these lines, for instance [79].

The model of the design process based on the theory of technical systems, as incorporated into design science, includes (by implication and explicit mention) many of the previously described methods [14, 44, 81, 130, 162, 213]. It is supported by some case studies that demonstrate the recommended process [29, 66, 119, 121, 122].

Other studies vital to design include psychology, application and control of mental processes, creativity [24, 52, 53, 89, 147, 166], management (of the design process [149], and of the organization within which design is performed), and many other interrelated topics. This includes cultural, sociological and regional attitudes about design work and about engineers [59].

Additional peripheral topics included in considerations of engineering design are aesthetics [36, 157, 201], modelling processes [27, 111, 199, 200, 201], conceptual considerations [81], analysis and optimization, and product design for assemblability [32].

Application of the theory of technical systems as starting point for theories of special branches and areas was the subject of section 12.1, which also showed that the presented insights are basic for rational developments in data processing for engineering design.

12.2.1 Design Science

Engineering design needs a large body of knowledge FOR and ABOUT the design process, to which we can apply the term Design Science [126, 127]. Design science comprises a collection (a system) of logically connected knowledge in the area of design. It contains a scientific study of the design activity in its context, including concepts of technical information and of design methodology.

Design science addresses the problem of determining and categorizing all regular phenomena of the systems to be designed, and of the design process. Design science is also concerned with deriving appropriate information in a form suitable for the designer's use, among others from the applied knowledge of the natural sciences (including information science and psychology), and from practical experience. Types of design science can be categorized according to their area of validity (general, vs. special for a particular area of expertise), and according to content (theoretical, vs. applied).

Any science passes through various stages of maturity, which may be summarized as [92]:

(1) Description of phenomena (natural history phase).
(2) Categorization in terms of apparently significant concepts.
(3) Ordered categorization whose pattern may be deemed a model (the evolutionary taxonomy or periodic table phase).
(4) Isolation and test of phenomena, with implied reproducibility by independent observers.
(5) Quantification (classical physics phase).

Even the first of these is commonly acknowledged as "a science" when applied to a traditional area of study. Only the last of these stages can lead to an "exact" science, if the quantification can occur with sufficient precision, repeatability and explainability, and if assumptions can be found to reduce the number of variables to be considered.

It is clear that a body of knowledge can be developed to describe the processes, procedures and products of engineering design. **Design science** in this sense will never reach the stage of an "exact" science. Like any other science, it can be developed in one of two ways:

- by the conventional empirical way of observing, describing, abstracting, generalizing, formulating guidelines, modelling, refining;
- by postulating a set of hypotheses, formulating a theory, modelling, refining, and only subsequently testing.

These approaches are not mutually exclusive. Both approaches should deliver a workable system, but the recommended result achieved by degrading from a

recognizable ideal should deliver better results. This circumstance should be applicable to a theory and science of design, as well as to any technical, industrial or management product or process.

The contents of Design Science can be structured in the form of a morphology [126], and as a **system of statements** (individual statements, and their relationships):

(a) whose parts and elements have certain **purposes,**
(b) that are **supported** (substantiated) in certain ways,
(c) that each illuminate only certain **aspects** of designing, and
(d) that are directed at a certain circle of **recipients.**

	Characteristic	State of Embodiment A	State of Embodiment B	State of Embodiment C
1	Methodological Category of Statement	Primarily descriptive (d–statement)	Primarily prescriptive (p–statement)	Normative (n–statement)
2	Empirical Support for Statement	Pre–scientific (practice – experiences)	Scientific singular "understanding"	Scientific statistical (inductive)
3	Intended Recipient of Statement	Novice, student	Teacher, researcher	Practitioner, designer
4	Aspects of Designing (interest filter)	The technical system (in usable state)	The design process itself	
5	Range of Objects as Subject of Statement	Universal: all artificial real and process systems	Technical real systems = TS	Branch–specific objects: machine systems, building systems,etc.
6	Declared Aims of Researcher	Automation of parts of the design process	Better empirical foundation for methodology	Others
	Any given statement of design science may be characterized by combining the appropriate states of embodiment from each characteristic.			

Fig. 12.3 Morphology of Statements on Design Science

Figure 12.3 shows this morphology of the statement system "design science". This morphological categorization of design-scientific statements brings design science into the third stage of maturity [92]. (The "states of embodiment" in this figure are interpreted as alternative solutions to the problem stated by the "characteristic"; the format of columns labelled (A), (B) and (C) is only convenient for referring to these states of embodiment, it has no other significance.)

If we now use the states of embodiment of characteristic 4 (technical systems, and design processes) as a first coordinate, and two states of characteristic 1 as a second coordinate level we obtain the following four basic classes of statements (see figure 12.4):

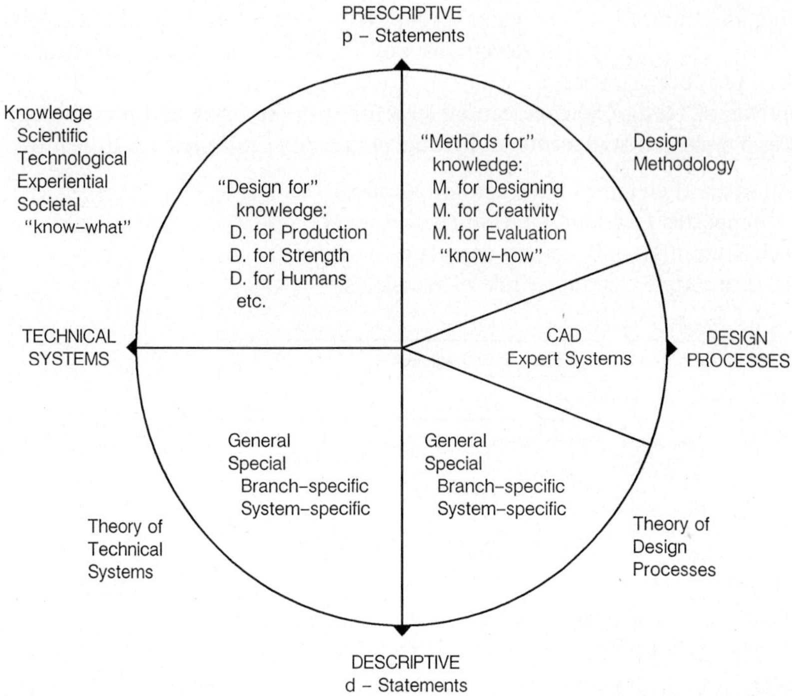

Fig. 12.4 Two Dimensions of Design Science

A. **Descriptive** statements about technical systems – the Theory of Technical Systems as presented in this book
B. **Prescriptive** statements about technical systems – the "specialist's domain- or branch-specific design knowledge" which contains the know-how about realizing a TS, and fulfilling functions and other requirements.
C. **Descriptive** statements about the design process – the Theory of Design Processes [112, 119].
D. **Prescriptive** statements about the design process – the "methodology of the design process", which shows ways in which the design process can be successfully performed in the socio-technical context [119].

The importance of working means (the designer's tools and equipment, see figure 12.2) is so great that it appears of advantage for the range of problems to add the following class of statements:

E. **Descriptive** and **prescriptive** statements about working means, particularly about application of computers: The term "working means" includes all equipment, appliances, devices and machines at the workplace which directly support designers in their problem-solving processes.

Design science, in its current state of development, is an attempt to provide a general, comprehensive theory of designing. It can be concretized for the processes of designing a variety of technical processes and systems, just as the theory of technical systems can be concretized (see section 12.1). Figure 12.5 indicates a hierarchy

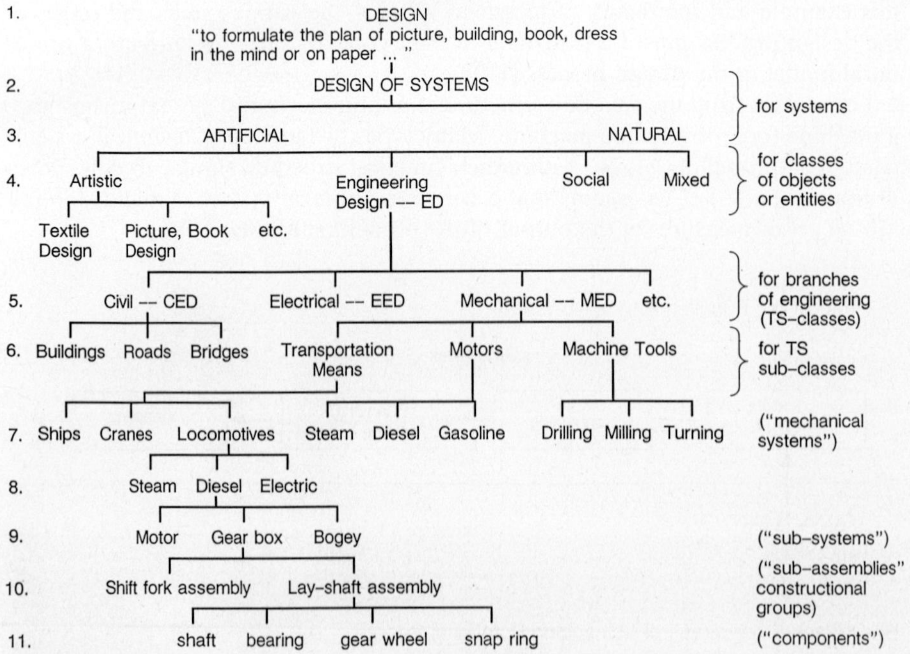

Fig. 12.5 The Systems Family - Hierarchy of Design Tasks

of systems to which design science can be applied. It should supply guidelines for improving the work of engineering designers, and thereby cover all aspects of the design process as indicated in figure 12.2.

Without limiting the designer's creative and intuitive abilities, the theory of technical systems is intended to lead the engineer teleologically to think in purposeful ways, to perceive large-scale connections, to understand and use holism as a principle, and to recognize generally those analogies and relationships that are important. These abilities are particularly meaningful for the design engineer. Wide distribution of knowledge about this theory among engineering designers is thus a desirable aim. Further details about the design process may be found in [119], including recommended procedures, steps, stages and methods derived from the theory of technical systems and design science.

12.2.2 Case Study: Designing a Worm-Gear Drive and Its Bearings

The following example demonstrates some of the difficulties of engineering design, even though it relates to a relatively simple subsystem. It also shows that a general theory of engineering design (based on design science including the theory of technical systems) can be concretized and applied to yield a special theory of designing machine elements [170], in this case a worm-gear speed-reducer.

The emphasis in this case study example lies in selecting and designing the bearings and bearing mountings of a worm-gear speed-reducer, a task at about levels 10 and 11 of figure 12.5. In order to understand the relationships between

this example and the theory of technical systems, the various steps and stages of the design process must be explained in their context, and in relation to a procedural model of the design process [119].

Let us start from the function structure of a larger technical system for instance a machine tool, or a textile machine. Many types or families of technical systems contain a "propelling block" within their function structure, similar to that shown in figure 12.6 A. Let us assume that a concrete technical system demands the following set of measures at the output of the propelling block:

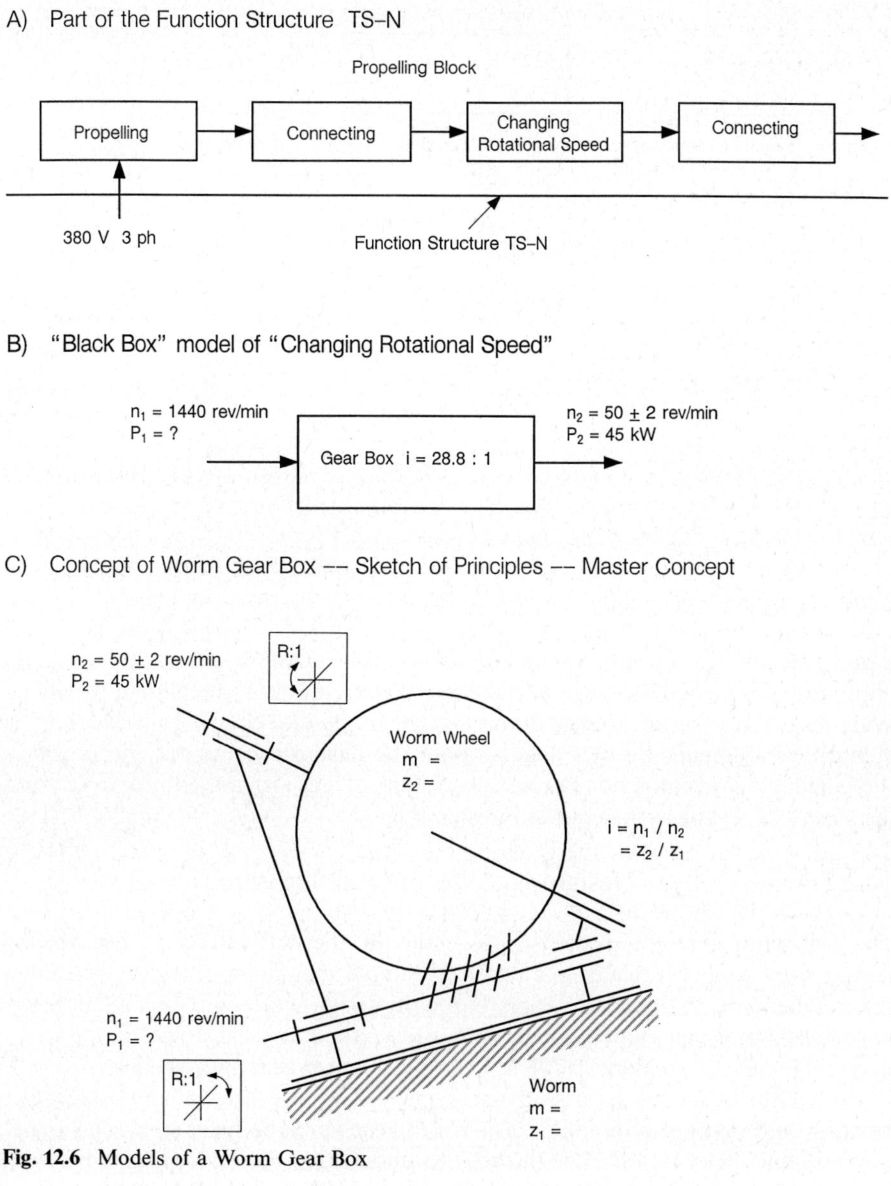

Fig. 12.6 Models of a Worm Gear Box

D) Organ Structure of Worm Gear Box –– Master Organ Structure

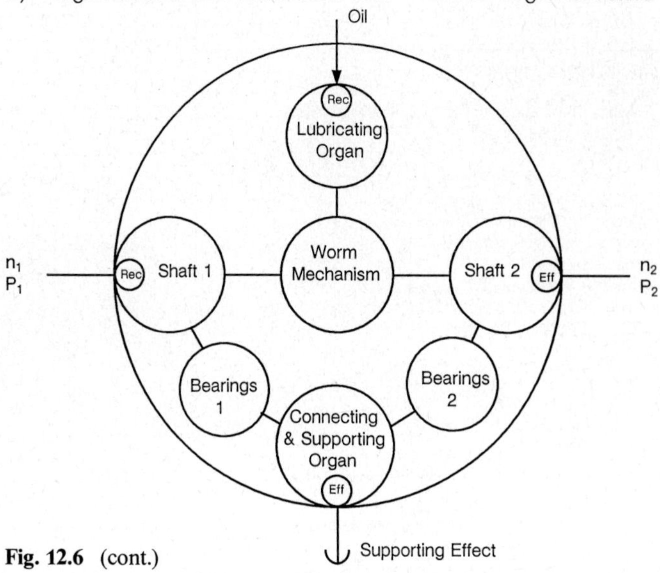

Fig. 12.6 (cont.)

rotational speed	50 ±2 rev/min
power	4.5 kW
shaft diameter	30 mm

These considerations take place at the level of the total system, namely in level A shown in figure 12.7. When designing the propelling block (which is a partial system of the total system) we move to the next lower level B. A design specification for this partial system is developed from the requirements, and some relevant duties of the sub-system are specified as shown in figure 12.7, level B. Developing a design specification is emphasized here, this is always a necessary step for the designer to understand the problem and its implications [119, 152].

Designing the partial system "Speed Reducer" takes place on the next lower hierarchical level, figure 12.7 C. As with every design problem, the list of requirements and design specification is considered and adapted, as the first step in the recommended methodical design procedure. More rapid progress can be made by using an existing design taxonomy, figure 12.8 A. This taxonomy fulfills a similar task to the survey parts of the published design catalogs, see for example [187]. Selection of the general types of speed reducer suitable for this project is assisted by the data table shown in figure 12.9. The selected worm gear box could basically be an existing product that is re-used for this different application - either from the drawings and data produced for a previously designed solution within the company, or as a purchased (bought out or out-sourced) item from a company that specializes in making gear boxes. In this example we will follow the path of producing a newly designed gear box.

In this case, even the design process for a new system turns out to be fairly simple, because we are dealing with a generally known artifact for which we possess basic sample solutions that we can term "masters", see figure 12.6 B, C and D.

Solution Level	Proced. Model No. <119>	Design Step	Result
Level A TS N (total system)	1	a) Set up design specification of TS N	
	2	b) Establish function structure	see Fig. 12.6A
	3–4	c) Establish values for entry to Level B	P = 4.5 kW n = 50 +_ 2 rev/min
Level B	1	a) Derive more detailed design specification from TS N	
Propuls. Block	2	b) Analyse functions of propulsion block	
TS DRIVE	3	c) Propel –– select electric motor (Note: other variants were not considered here)	P = 5.5 kW (assumed = 0.8) n = 1440 rev/min
		d) Select connections	rigid flange couplings (other variants not considered)
		e) Change rotational speed	see Level C
Level C TS GEAR BOX	1	a) Derive design specification from TS DRIVE	To occupy small space, rotate in one direction, shaft position arbitrary
	2.1	b) Represent as black box	see Fig. 12.6B
	2.2	c) Establish principle of action	
	3.2	d) Establish classes of function carriers (degenerates to selection from taxonomy, Figs. 12.8 & 12.9)	Worm gear box
	3.3 3.5	e) Establish organ structure –– master concept principle	Fig. 12.6C & D
	4	f) Establish preliminary layout	Master layout (not shown here)
	4.1	g) Establish leading dimensions, i.e. calculate worm gear (theoretically at level D)	
	4.2 – 4.5	h) Establish organs –– lay out worm mechanism, lubrication, connections, etc.	Fig. 12.10A

Fig. 12.7 Procedural Plan for Designing a Worm Gear Box

Progress in this design process can be compared with figure 5.11, and the models used in developing the solution can be referred to figure 5.4.

The structural elements and their relationships are thus automatically defined. Figure 12.7C shows an interesting aspect of representation, which is especially useful in the conceptual phases of engineering design. When thinking about the required functions of a technical system, it is usually not yet possible to allocate locations of components in three-dimensional space. This figure shows some similarity to the "surrealist" paintings by Pablo Picasso, the system is simultaneously viewed from many directions. The relevant parts are juxtaposed and superimposed to give a diagram of connectivity, with little implication of spatial relationships.

Design work continues on the lower hierarchical level D (or even E) of figure 12.7. The basis for such a hierarchical sub-division is a suitable choice of system

Solution Level	Proced. Model No. <119>	Design Step	Result
Level D	1	a) Define problem from design specification for TS GEAR BOX and layout Fig. 12.10A	
| TS BEARING SYSTEM 1	4.1	b) Lay out shaft	
		- establish loading regime	
		- establish minimum size at critical locations	
(then TS bearing system 2)		- establish form -- geometry considering assembly, costs	see Fig. 12.10
		c) Select bearing (one at a time)	
		- establish bearing concept of one system (e.g. wheel shaft)	2 support bearings
		- design specification	
		- assume bearing type	single-row ball bearing
		- assume bearing size (from shaft)	
		- check bearing selection -- load capacity, speed, life, temperature, etc.	
		- calculate bearing lubrication	
		d) Bearing -- lubrication	Oil from worm gear box
		e) Bearing closure and sealing	Fig. 12.10B
Level E		......	......

Fig. 12.7 (cont.)

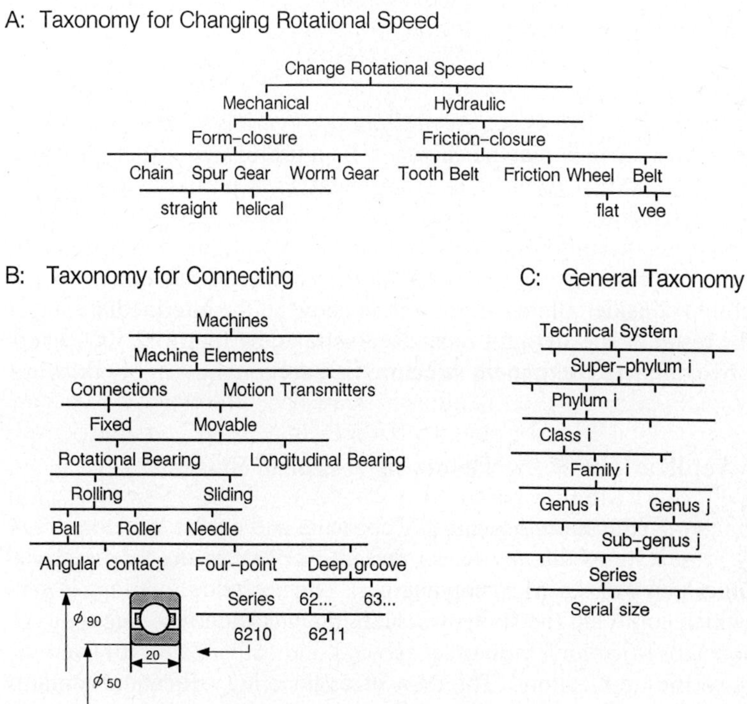

Fig. 12.8 TS Taxonomy (selected sections)

Parameter matrix for the function "transmitting torque from one shaft to another"								
Function carrier	Important parameters or characteristics							
Family of TS that can fulfill the function	Possible shaft arrange-ments	Maximum ratio i_{max}	Maximum power P_{max} kW	Maximum surface speed v_{max} m/s	Average efficiency η	Average working life t_L	Relative cost guide	etc.
Spur gears -- straight tooth	||	4 (7)	50.10^3	15 (100)	0.98	10^5		
Spur gears -- helical tooth	|| |•	7 (10)	50.10^3	25 (150)	0.97	10^5		
Worm gears	|•	4 (100)	50 (200)	15	0.4–0.7	10^3		
Friction wheels	||	7 (15)	100	15 – 20	0.96	2.10^3	Guideline values may be found for specific cases	
Chain drives	||	8 (10)	100 (3500)	40	0.9–0.97	$2–10.10^3$		
Flat belts	|| (|•)	5 (20)	100 (3500)	35 (100)	0.88–0.95	Large variations		
Vee belts	|| (|/)	7 (10)	100 (1500)	60	0.96	Large variations		
Rope drives	|| (|•)	5						
Tooth-belts	||							
etc.								

Fig. 12.9 Parameter Matrix for the Function "Transmitting Torque from One Shaft to Another"

boundaries. If, for instance, we choose to declare the bearing system as a partial system of the gear box – which may be shown to be a useful step – then the bearing itself, as a constructional element (a component) is a partial system of the bearing system, and selecting the bearing takes place on the hierarchical level E.

The general progress, starting from the situation shown in figure 12.10 A, can be followed from the descriptions of figure 12.7. In the context of this book it is not possible or useful to consider all the details and to show all the intermediate stages and models. The result of this design process is illustrated by figure 12.10 B, a representation of the complete component structure in a sectional assembly drawing.

12.3 Further Applications of the Theory of Technical Systems

Regarding engineering education, present-day curricula and course structures contain hardly any presentations that give a connected overview of the total material (e.g. for mechanical, civil or electrical engineering). The necessary topology is generally missing which could aid the student in learning and understanding the contexts and relationships (a) among individual subjects and courses, and (b) from the curriculum to real-life applications. The area of engineering education contains many problems, and is subject to extensive discussions [11, 62, 70, 113, 123].

A) Problem

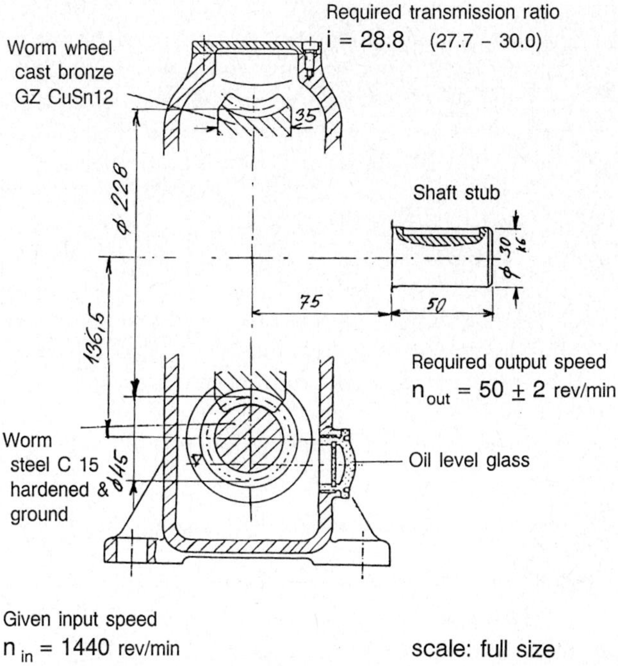

B) Result

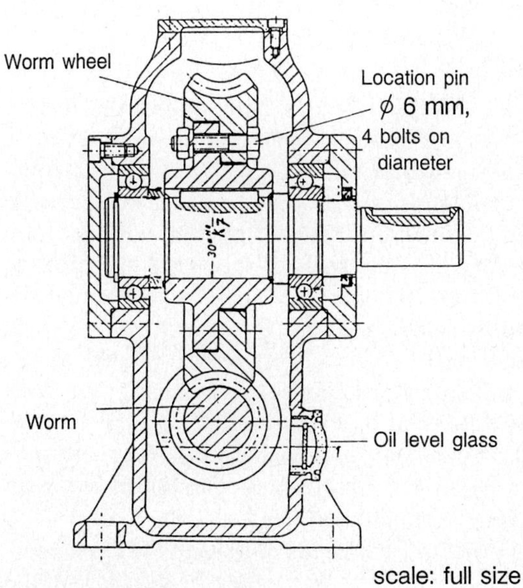

Worm -- number of starts	$z_1 = 2$
Worm wheel -- number of teeth	$z_2 = 57$
Actual transmission ratio	$i = 28.5$
Module	$m = 4$
Efficiency	$\eta = 0.8$
Power	$P_2 = 4.4$ kW

Fig. 12.10 Worm Gear Box

The theory of technical systems, when concretized for a particular branch of engineering (e.g. as the "theory of machine systems" for mechanical engineering), can perform the function of providing a connecting topology of engineering in its socio-technical context, because it treats all basic aspects from a sufficiently high level of abstraction, and offers adequate orientation about the area of learning (and about the whole of engineering in general) by its taxonomy. All courses, laboratory and tutorial presentations can be shown in their context with reference to this theory, and the educational objectives can be explained [63, 125]. It is expected that such an overview will encourage transferring knowledge between individual elements of the system of engineering education, and to assist students in understanding many of the course contents. This latter aspect should also cover the procedural problems of engineering design, roles of creativity, and communication (written and oral reporting), and provide some added motivation for learning. Various case studies have been prepared to give examples and opportunities for practical exercises [29, 66, 121, 122].

The theory of technical systems can fulfill the following functions in engineering education:

- Introduce students to engineering in general, and to the relevant specialized area.
- Render the whole educational system more transparent, and explain the place and purpose of individual subjects: for instance showing thermodynamics as a special theory of processes that contains the knowledge needed for the technical processes of thermal machinery, or showing mechanics of materials (strength) as the general study of a particular property of technical systems - resistance to load and failure - which serves to establish dimensions for components.
- Offer a survey of the range of problems in the technical area, whereby those topics that are missing in that area can be clearly recognized.
- Serve to summarize and recapitulate the transmitted subject matter, and emphasize their context.

The Theory of Technical Systems as presented in this book shows the overall structure of knowledge about processes and systems, including the roles of information, management, goal-setting and realizing, and the social and physical environment. The effects that act on the operands of a process, and that the technical systems produce, are defined as the TS functions. Their "internal workings" are described by their modes of action, and it is the duty of the "hard" sciences to describe and model the principles of these actions. The "hard" sciences and the social and humanistic sciences are thus subsumed into the Theory of Technical Systems. This theory is therefore the ideal "advanced organizer" for explaining technology and its role in society, and to give fundamental reasons for studying the subjects of an engineering curriculum [125].

A curriculum should be built in a rational way, and engineering design can provide the backbone and core to the sequence and individual courses that make up the curriculum [63]. Curriculum designers should consider:

- the activity, the teaching/learning process - the needs of students when preparing themselves for future practical activity, and the needs of society for the expertise;

- the information, the operand of the teaching/learning process - facts, phenomena, models, attitudes, motivation, feelings, etc., but also the types of informational input, and the desired outcome in the form of engineering abilities of the graduating students;
- the students, academic and technical staff, as operators of the process;
- tools and aids used as operators - lectures, seminars, laboratories, projects, assignments, problems, tutorials, case studies, course outlines and operating philosophies, coordination of subject matter and contextual experiences, but also computers, drafting equipment, textbooks, rooms, etc.;
- the environment as operator - the university or college, the city, the economic and cultural environment, including the needs for evaluation and assessment of the student's knowledge and abilities, but also of the staff members and their working conditions and management.

In order to learn to design (but only as a background to designing good products), students must learn a wide variety of knowledge, skills, abilities and attitudes. They must study many phenomena, incorporate many concepts into their mental framework, be familiar with a multitude of facts, descriptions and interpretations, and learn many theories. They must also discover (frequently for themselves) the relationships among these varied items, and how to manipulate them to enable their synthesis (compare [62]). It is only by internal juxtapositions that creative (novel) thoughts emerge.

In fact, all the knowledge of the sciences, of technology, sociology, psychology (and many more) serve only as heuristics and guidelines for real engineering work [142]. Students therefore need to discover when to try to be "exact" and when to use coarse approximations and order-of-magnitude studies.

This is not all, they must also learn HOW to design, the aspects of processes and procedures. All activities have some procedural content, and this can be explained to the students by some means. Activities also have speculative, judgemental, analytic and communications content that fit together into a good problem-solving procedure (almost an "algorithm"). Learning really only takes place by the students "doing something", namely practicing some useful procedures on some typical problems, and abstracting from that experience to generalize into a personal procedure that works for the student in all situations. Doing can range from purely routine to highly creative, using a suitable combination of verbal, graphical and symbolic tools, and students must be exposed to this wide range.

They also need to develop the courage to explore a wide range of candidate solutions before converging to one or two to be taken into the next more detailed stage. They need to iterate (go back to revise the work of a previous stage), and at times back-track to alter a previously-made decision. They need to separate their mental processes to make each one more effective. And they need to see a continuous (but by no means steady) progression from an abstract statement of needs, through a full specification for manufacture, to the working hardware, including all the pitfalls, frustrations, needs for additional knowledge, and negotiations to realize an economically viable solution for the final market.

The need for design education is to combine explanation and practice, to provide education and training, to stimulate motivation and interest, to incorporate

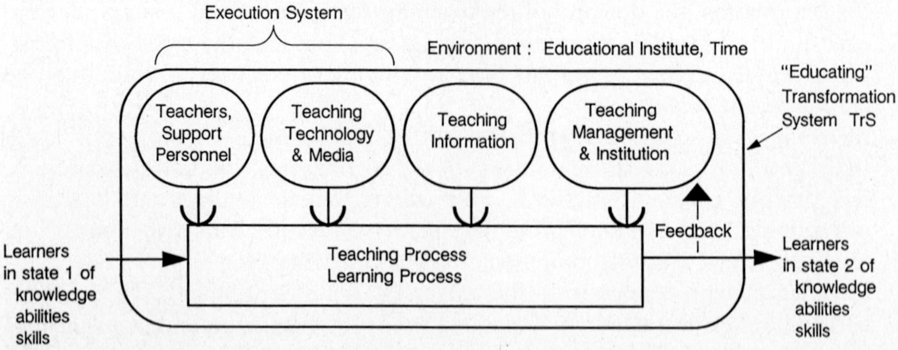

Fig. 12.11 Model of the Teaching/Learning Process

professionalism and ethics, and to expose some of the hard "facts" of economic life such as the needs for financing and returns on investment, and needs to accommodate ease of manufacture and servicing. This is liberal education, not in separate unrelated packages, but in a holistic overview.

Teaching (and learning) is also a process, and it is possible to establish a model of teaching similar to that of the socio-technical transformation process, see figure 12.11. The elements or partial systems of this process are:

- the student, whose state of knowledge, abilities, skills and experience is to be changed;
- teaching goals – formulation of the "ideal" output of the process;
- the array of optional and compulsory subject matter open to the student, which constitutes the formal content of the curriculum;
- the teaching system, namely performing the process of teaching, including the teacher and the means (equipment and educational media) used;
- information system for teaching, consisting of the knowledge that is to be taught, and the knowledge about how to teach (pedagogy and methods);
- management system that sets the goals, controls the procedures, and performs the functions of administration, and organizing to achieve the goals;
- the surroundings.

The theory of technical systems is also usable in general education. It can fill the existing gaps, and help to explain to everyone, even the technologically ignorant, the roles of engineering and technical systems in society, thus contributing to a better understanding of technical matters. The form of presentation must, of course, be adapted to the audience.

The history of technology in society can also be enriched by applying the theory of technical systems. The processes of development of technical systems can be better explained if the changes in design characteristics are consistently presented as dependent on the factors of the environment, including social, economic, technical, scientific and cultural aspects.

A coordinated application of these insights can make an important contribution to attaining a better relationship between humans (including engineers) and technology.

12.4 Relationship of TTS to Other Knowledge

Many related areas have received attention. The references quoted in this book give only a limited selection of those available in the literature. They can be classified into topic groups as follows:

Adjustment - [101]
Aesthetics - [36, 157, 201]
Alternatives - [25, 26]
Bibliography - [114]
Case examples - [29, 66, 121, 122]
Causality - [159, 172, 173]
Checking and verifying - [48]
Computer aids - [43, 56]
Codes of practice - [12–23]
Conceptualizing - [81]
Conferences - [68, 93, 116, 117, 120, 124]
Cost - [22, 23, 72, 73, 79] - see also economics
Cost-benefit - [83, 148]
Creativity - [24, 52, 53, 89, 147, 166]
Critical path method, PERT - [47, 96, 149, 197]
Culture - [59, 159, 168, 169]
Cybernetics - [35, 39, 135, 139, 151]
Decisions - [20, 26, 160, 176]
Design assurance, design audit - [48]
Design catalogs - [15, 186, 187]
Design characteristics, properties - [31]
Design, conceptual - [81]
Design education - [11, 62, 63, 70, 113, 123, 125]
Design for assembly - [32]
Design for economics - [21]
Design, general - [14, 37, 58, 102, 118, 142, 156, 160, 162, 210]
Design methods, history - [37, 44, 58, 75, 90, 99, 102, 133, 146]
Design methods, individual - [14, 44, 81, 130, 162, 214]
Design science - [65, 67, 86, 92, 110, 126, 127, 207]
Design, systematic method - [13, 14, 15, 40, 90, 98, 100, 119, 140, 143, 167, 170, 180, 182, 210]
Design systems theory - [138, 160, 162, 202, 212, 213]
Design theory - [28, 44, 64, 65, 67, 74, 99, 100, 108, 112, 119, 133, 156, 167]
Designer (general) - [76, 77, 168, 181, 194]
Detail design, machine design - [58, 85, 129, 133, 150, 167, 188, 193, 205]
Developments in time - [107]
Development (TS) - [8, 204]
Dictionaries - [3, 9, 135, 191]
Dimensions & units - [42, 60, 61, 155, 174]
Economics - [18, 19, 20, 22, 78]
Encyclopedias - [1, 2, 4, 5, 7]

Ergonomics – [21, 71, 131, 134, 209]
Evaluation – [132, 175, 204]
Expert systems – [56]
Failures – [164, 183, 208] – see also reliability
Form-design – [17, 133, 156, 205]
Future – [141, 161, 168, 169, 194]
Heuristics – [5, 142]
Historic – [33, 34, 178, 179]
Human factors – [134]
Idea generation – [24, 52, 53, 89, 102, 147, 166] – see also creativity
Information – [6, 187]
Invention – [95]
Learning curve – [78, 203]
Management – [47, 55, 88, 96, 149, 197, 198, 211] – see also motivation
Market suitability and requirements – [46, 152, 168]
Measurement of engineering activity – [8]
Modelling – [27, 111, 171, 199, 200, 201]
Morphology – [119, 214]
Motivation & sociology – [51, 103, 104, 105, 153, 154, 211]
Optimization – [50, 80, 129, 145, 160, 177, 196]
Parts classification – [106, 165]
Product development – [46, 204]
Product planning – [12]
Reliability, maintainability – [91, 128, 175, 195, 206]
Representing – [27, 38, 45, 57, 82, 87, 111, 199, 200] – see also modelling
Socio-technical system – [54, 184, 185]
Specifications – [152]
Standards – [10]
Statistics – [41, 49, 97, 163, 189]
Systems engineering – [98, 137, 138] – see also cybernetics, systems theory
Systems theory – [136, 137, 138, 202, 212, 213] – see also cybernetics
Terminology – [16, 115]
Testing – [190]
Theory of machines, modelling of mechanisms – [192]
Theory of TS – [30, 69, 109, 184, 185]
Tolerancing – [43]
Value analysis – [22, 84, 94, 144, 158]

These references demonstrate the broad concepts and developments necessary to understand and practice engineering. More extensive bibliographies are available (e.g. [114]). The relationships between these areas are complex. Design science, represented mainly by the theory of technical systems and the theory of design processes [112, 119], together with the theory of socio-technical systems [54, 184, 185] shows how these detail fields relate to one another. The theory of technical systems presented in this book is therefore a powerful tool for understanding many human situations.

Appendix A: Summary of Statements and Propositions from Chapters

Chapter 3: Transformation Systems – Statements

Statement 3.1 Human society has many needs, and these depend on that society's system of aims. Each individual aim (as an element of the system of aims) is a requirement for a particular object in a particular state. The objects can be regarded as the means of achieving the aims. If a desired state does not exist, then a new aim is evoked with a need to cause a transformation (to generate a state) that permits achieving the primary aim.

Statement 3.2 The changes of state are termed transformations, and the object that is being transformed is termed operand.

Statement 3.3 A transformation is an artificial process in which the changes are achieved by using natural phenomena.

Statement 3.4 The transformation is required (evoked) either because an unsatisfactory input state of the operand exists (when the choices concern only the transformation), or by defining an output state for the operand such that it provides the desired means to fulfill the aim (when both the input state and the transformation may be chosen).

Statement 3.5 Classes of operands in the transformation system are:

- biological systems – living matter, including humans, animals, plants, bacteria, vira, etc.
- materials – gases, liquids, solids, plasmas, etc.
- energy
- information – data, commands, news, signals, etc.

Statement 3.6 The transformation is performed by a technology that defines a system of partial transformations. In general, a number of technologies based on different technological principles are available to perform any transformation.

Statement 3.7 The transformations and partial transformations are realized by certain effects being exerted on the operand. Effects produced by the operators are the means of achieving the transformation. They are classifiable into material, energy and information (data and command) types.

Statement 3.8 The effects are delivered by various operator systems - humans, technical systems, information systems, management systems, and the active environment.

Statement 3.9 The transformation system is defined as the set of all elements that participate in the transformation. The model of a general transformation system is shown in figure 3.2.

Statement 3.10 The effects are produced by the acting operators, by means of action processes and action chains that transform the inputs of the operators into their outputs.

Statement 3.11 A causal relationship exists between the transformations of the operand (as consequence) and the action processes (as cause).

Chapter 4: Technical Process - Statements

PROPOSITION 4.1

The transformation (and thus the behavior of the system or process) $Od^1 \rightarrow Od^2$ may be realized by various structures of processes (and therefore by various technologies).

Statement 4.1 A technical process (TP) is a special form of transformation process in which technical systems are used by humans as artificial tools. Consequently, all statements about transformation processes as listed in section 3.4 are true for technical processes.

Statement 4.2 The structure of a TP consists of a set of partial processes or operations, and a set of intermediate states of the operand.

Statement 4.3 Partial structures can be produced for the three phases of preparing, executing and finishing (closure or conclusion) to a TP.

Statement 4.4 A process in which the technical system itself is the operand (instead of a separate operand that is subject to transformation) is only acceptable in exceptional cases, and for pragmatic reasons.

Statement 4.5 Each transformation process (the work process that transforms the operand) requires for its performance an additional set of processes: auxiliary, regulating and controlling, material and energy converting (propelling), and connecting and supporting processes. The need for these processes is evoked by the main working process. The representation of a TP includes (by convention and agreement) all these evoked processes.

Statement 4.6 When considering a TP, its environment (and space) and the time must also be considered and specified.

Statement 4.7 Secondary inputs and outputs (including disturbances, waste and pollutants) belong to the TP-system and must be considered.

Statement 4.8 Even though technical systems are the most important operators of a TP, all other operators must be investigated.

Chapter 5: Technical Systems - Statements

Statement 5.1 The task (aim, purpose) of technical systems consists of exerting particular effects on the operands in the technical process. Consequently, the TS must have the ability to exert these particular effects.

Statement 5.2 The effects are exerted from a system of action sites - effectors - (at the boundary of a technical system) under certain action conditions (compare figure 5.7).

Statement 5.3 The effect is achieved as the output of an action process (through an action chain) internal to the TS, in which the input measures are converted into the effects (output measures) of that TS. The action process is a direct consequence of the structure of the TS, the term "mode of action" of the technical system is used to describe this relationship. The action process takes place after it has been initiated at the desired moment.

Statement 5.4 The action process (and the mode of action) is characterized by the TS-internal conversion of inputs of material, energy and information. These internal conversions are termed (technical) functions, and are linked by their relationships in the function structure of the TS.

Statement 5.5 The (technical) function describes the capability of the TS to transform an input quantity into an output quantity of different measure and/or form by natural phenomena that are deliberately caused to take place.

Statement 5.6 The input quantities to a technical system are in the forms of materials, energy and/or information.

Statement 5.7 Technical systems exist during their origination, operation and disposal phases in various roles and states, according to the transformation systems within which they belong at that time.

Statement 5.8 In general, technical systems possess specific relationships to all other elements of the appropriate transformation system. Apart from the desired inputs and outputs, various other (undesired) inputs and outputs exist that are termed disturbances.

Statement 5.9 The functions (and function structures) are realized by organs (and organ structures) using certain action principles. Organs are the means employed to fulfill the aims of the functions. Each function (function structure) can be realized by several existing action principles.

Statement 5.10 The organs (and organ structures) are realized by constructional elements (and component structures). The component structure realizes also all types of properties, as will be discussed in chapter 7. Each organ (and organ structure) can be realized by several available component structures.

Statement 5.11 Each of the TS-structures depends on the state of embodiment of certain design characteristics. This is the cause for the alternative TS-structures that are possible.

Statement 5.12 Each function, each organ or constructional element may be resolved into a system of partial functions, partial organs or partial constructional elements, resulting from the hierarchical character of technical systems.

Chapter 6: TS Systematics – Statements

Statement 6.1 Technical systems may be classified according to various characteristics that they have in common, seen from different points of view. The set of all technical systems can be divided into a number of different sub-sets according to various criteria. The resulting categories can serve various purposes, such as creating order, overview, cataloging, evaluating, characterizing, understanding, comparing, etc.

Chapter 7: TS Properties – Statements

The category of properties is of central significance to the theory of technical system, the properties determine the object and its state of existence.

PROPOSITION 7.1

A technical system must be capable of performing the desired functions to fulfill its requirements, and must also possess certain other properties in an appropriate measure and form.

A technical system possesses certain properties, for instance with respect to minimum life or corrosion resistance, even if the requirements for these properties have not been explicitly stated.

Each technical system possesses all types of property, but their state of embodiment and measure (magnitude, value) can be present in different levels. The sum of these measures determines the total *quality* (both of design and of conformity) of the technical system.

PROPOSITION 7.2
The behavior of a technical system is determined by the structure of that system.

PROPOSITION 7.3
A relatively closed system with a given structure exhibits a single behavior, determined by its structure.

PROPOSITION 7.4
The observed behavior does not uniquely determine the structure that caused it. The same behavior can be realized by a number of different structures.

PROPOSITION 7.5
Every function (or behavior) is a system that contains as its elements a set of partial functions (or partial behaviors), and that is decomposable to elementary functions (or elementary behaviors). At each level of complication (from total function to elementary function) we can work with various degrees of abstraction (ranging from an abstract function to the actual behavior of a concrete technical system).

The particular functions "connecting" and "separating" are assumed to be elementary functions at a high level of abstraction. They combine a series of functions from various steps in the hierarchy of abstractions, and various stages of imposed conditions (see figure 7.3, and compare figure 5.5)

PROPOSITION 7.6
All types of properties are achieved by means of the elementary design properties (structure, form, size, material, surface, tolerances, manufacturing methods)

PROPOSITION 7.7
Every technical system (machine system) can be decomposed into partial systems (machines, constructional groups, sub-assemblies, constructional elements) which in their prescribed combination fulfill the partial and elementary functions of the TS.

PROPOSITION 7.8
The behavior of a technical system depends not only on the sum of behaviors of the elements, but also on the coupling relationships between these elements.

PROPOSITION 7.9
The basic factors on which the value of a realized technical system depends are: the abilities of the design team (see figure 7.22), design time (see figure 7.23), and the number of improvements.

Statement 7.1 Technical systems must possess their purpose function, the capability of action, and the capability of exerting suitable effects as their main output.

They must also possess certain well-defined measures or states of embodiment of a finite set of properties which make technical systems suitable for use. Some of these properties provide direct support for the capability of action, others form additional aspects of the product.

Statement 7.2 A full listing and survey of the properties may best be obtained by considering the classes of properties and aspects (see figures 7.2 and 7.16)

Statement 7.3 Every technical system possesses all types of properties, whether these are consciously planned according to the requirements, or not.

Statement 7.4 The starting point of technical systems are the requirements that determine the properties of the future TS.

Statement 7.5 Properties may be categorized according to various viewpoints. Important classes are the external (usage) properties, and the internal (design) properties.

Statement 7.6 External properties depend on internal design characteristics and properties (see figure 7.16).

Chapter 8: TS Evaluation – Statements

Statement 8.1 The total verdict (or partial verdict) about the quality (or total value, or partial value) of a technical system may be characterized by a (vectorial) sum of measures of selected properties.

Statement 8.2 The measures of properties (as their states of embodiment) should be determined by the most objective measuring scale available (problem of quantifying the properties).

Statement 8.3 The selected properties are termed criteria; selecting them is one of the most important factors for the quality of the evaluation (or comparison).

Statement 8.4 Processing of the measures of individual criteria to a total value can be accomplished by various methods. The selected method is also one of the important factors for the quality of the evaluation.

Statement 8.5 A correct determination of the measures of properties in some phases of the life of a TS is a difficult problem. An objective determination or estimation of the measures supports the quality of the evaluation.

Chapter 9: Representation of Technical Systems – Statements

Statement 9.1 Representations of technical systems describe models of TS with their design properties and design characteristics.

Statement 9.2 According to the classes and numbers of the described design properties, representations can be relatively abstract (when abstracting from some design characteristics) or concrete (and usually more complete), compare also figures 9.1 and 5.11. In other words: The model of the technical system described in the representation can be more abstract or more concrete.

Statement 9.3 Every TS model can be represented by various methods and techniques of representation.

Statement 9.4 Every representation serves one (or more) certain purpose, and the kind of representation must be suited to that purpose.

Chapter 10: Origination of Technical Systems – Statements

Statement 10.1 The life of a technical system is a process system that can be defined as a system of partial processes and their relationships.

Statement 10.2 The four characteristic basic phases of a TS life are: origination, distribution, working operation, and liquidation (see figure 10.1).

Statement 10.3 Every technical system must pass through all these phases, and fulfill the requirements of each phase, i.e. it must possess appropriate properties (see statement 7.1).

Statement 10.4 The number and scope of the smaller stages within the four basic phases depends on the following factors: complexity of the technical system (figure 6.2), originality of design (section 6.7), type of production (section 6.8), and customer (see figure 7.21). According to the type and traditions of the organization executing the phases, they can be very varied in their content and conformation.

Statement 10.5 Decisions on the total value of a technical system are mainly made in the design phase.

Chapter 11: TS Developments in Time – Statements

Statement 11.1 Technical long-term developments should support and ensure humanitarian goals of society. All engineers, technologists and technicians should do as much as is in their power to ensure that technical developments serve these societal goals, and are not misused in the interests of individuals or parasitic groups.

Statement 11.2 Developments in technical systems in the course of time consists of a chain of elementary changes in the properties of technical systems. The cumulative consequences of many small steps can appear as a quantum leap – a technical break-through.

Statement 11.3 The speed of long-term developments of TP and TS is increasing, whereby:

- shorter time from idea to realized TS is demanded;
- shorter effective lives of technical systems become acceptable;
- requirements for certain properties of technical systems change their trends.

Statement 11.4 The long-term developments of technical systems are influenced by three main factors:

- technical capability
- economic potential,
- motivation towards such developments.

Other important effects can come from the raw material situation of a country, and the operators of the design process.

Appendix B: Terminology

B.1 Discussion of Basic Terminology

When we formulate thoughts to communicate them in normal language, we usually use terms that we have acquired by experience. This informal method is generally unsuitable when we are trying to build up a scientific discipline, because for that purpose the means of expression must have a clearly delimited, preferably unique and compelling meaning suitable for the field of knowledge. Some terms are only used in one scientific field, e.g. buckling resistance. Other terms are available in normal language, and usually have imprecise or multiple interpretations. In the context of a technical or scientific field, they are either used with more precisely defined content (e.g. system, tension), or have completely different meaning (e.g. sleeve, bush, tooth, ram).

The activity of defining a terminology is not a popular one, and reading such definitions usually meets with resistance from readers. We cannot avoid this step for the task of presenting the theory of technical systems, especially if we wish to reduce the possibilities of misunderstandings.

A different problem arises from making suitable choices of terms for technical expressions. This decision is difficult, as was demonstrated in chapter 1 with respect to an expression for the general abstract machine. Criticisms come easily, especially from those people who are working on similar topics and have chosen a different term to express a similar content. This danger can only be avoided if the differing opinions are confronted and discussed at an early stage, and some measure of unity can be reached. This is particularly so in the fields of engineering and general technology, where a rapid development from empiricism to applicable theory has taken place. There is thus a tradition of intuitively accepting and retaining existing terms, or coining terms from loose analogies, almost without question, and without exactly defining them. For instance, the terms "machine" and "engine" as stem for a whole series of other terms have vastly differing content, depending on engineering field, time and place – "engine" can be a prime mover for automotive machinery, or a device for providing regular scribed rulings on a scale, or a host of other items of equipment.

The terminological difficulties are reinforced by the different contents and meanings of similar words in different languages. We discussed some differences between German and English in chapter 1 with respect to "Technik" and "technology". The German "Konstrukteur" is only related to the English "designer", the two terms do not coincide. Other discussions of similar difficulties are contained in section B.3 of this appendix.

It is worth mentioning in passing that agreement about words is still missing in some of the basic sciences. This is true of cybernetics and systems theory, both of which supply important points of departure for the theory of technical systems. Such lack of general agreement forces us to survey some elementary terminology, instead of simply referring the reader to other literature.

The selection of technical terms for this book was based on the following principles:

- largely use of existing expressions in their conventional meaning, but with more precise definitions;
- connections with general sciences, e.g. mathematics;
- use, wherever possible of an international set of words to help attain international understanding;
- consistent use of the same word for the same meaning, which may render the subject matter "dry" for those used to the variety of words appearing in English literature.

Some other works in the literature have helped in the search for terminology and definitions [100, 115, 185, 187]. We aim towards precision and comprehensive understanding, if the free flow and beauty of language *per se* suffers to some extent, so be it.

Various terms are accompanied by recommended symbols that are intended to help with further formalization, and as a form of short-hand notation to save work for the engineer. It is up to the reader to use or ignore these possibilities. Only a few such symbols are used within the text of this book, mainly to avoid continuous repetition of terms where the context is likely to be clear. These are collected as a list of symbols in the preface section. "Decoding" should thus be easier where it cannot be avoided.

Definition of the terminology takes place in two stages. This appendix (in the following two sections) contains the definitions and discussions of a selection of basic terms that should be familiar to most readers. The special terminology for the theory of technical systems is presented within the twelve chapters, as their themes demand it.

The basic terms discussed in section B.2 have been selected according to the sequence in which they are likely to be needed for defining the special terms of our theory of technical systems. With the aim of easier orientation, the basic terms have been collected into suitable thematic regions. The relationships between these regions have been used to decide on their sequencing – the definition of "set" is used to help define, and therefore precedes, the term "system".

This selection can not contain all of the basic terms that could be or are used. For instance, formal logic and the logic of predicates has not been included here. Such regions of thought have not reached a sufficiently wide audience, and reading strings of logical symbols is likely to prove tedious for many readers. Some noteworthy attempts have been made in the literature, for instance [185, 212, 213].

Section B.3 contains some informal discussions and polemics about the reasons for choosing some words or expressions in preference to others.

B.2 Formal Definitions

SET A *set* is a collection of objects, thoughts or opinions (its elements) to a conceptual whole. According to the number of elements present in the set, we distinguish null-sets (no elements), finite sets, and infinite sets.

If **X** is an element of the set **S**, the usual notation (see also figure B.1) is:

$$X \in S$$

Two sets are *equivalent* when each element of the set **S** maps uniquely to one element of the set **T**, and this mapping is mutual:

$$S = T$$

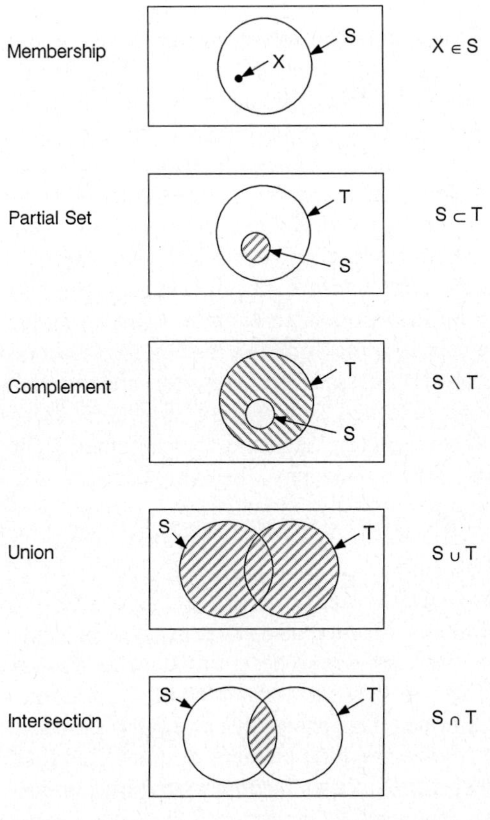

Fig. B.1 Operations on Sets

A *partial set* exists when all elements of the set **S** are contained in the non-equivalent set **T**:

$$S \subset T$$

The total of elements **T** minus **S**, i.e. all elements of set **S** that are not contained in set **T**, is termed its *complement*:

$$S \setminus T$$

The *union* contains all elements that are contained in either set **S** or **T** or simultaneously in both sets **S** and **T** :

$$S \cup T$$

The *intersection* contains all elements that occur in both set **S** and set **T**:

$$S \cap T$$

SYSTEM (S)

A *system* is a finite set of *elements* collected to form a whole under certain well-defined rules, whereby certain definite *relationships* exist between the elements, and to its environment.

It is possible that a system can contain isolated elements (i.e. elements with no relationships to others – relationships equal zero), or isolated groups of elements (i.e. the group has no relationships to other elements or groups within the set).

Both the terms "element" and "system" are *relative.* An element can also be regarded as a system, and a system can be regarded as an element within a larger system. Systems are hierarchical. For instance, a machine is a system that may be composed of machine elements (including for instance brake, motor, speed-reducer, etc.). Each machine element may be regarded as a system, and consists of smaller machine elements (e.g. rolling bearings) and separate components (shaft, cover, housing). In the other direction of this hierarchy, the whole machine is likely to be an element in a factory or plant.

A system can be divided into partial systems (sub-systems) of various degrees of *complexity.* Each system can therefore be studied at different *levels of discrimination.* These levels depend not only on the specific field of engineering (or science, or sociology), but also depend on the abilities of our sensing organs, and the technical means available to assist them. For the purposes of designing a technical system, we do not usually sub-divide a system to the levels of molecules and atoms, except to understand the behavior of materials, their properties and failure modes. The usual lowest level of discrimination is the machine element or component, or the system of geometric body shapes from which it may be constituted for a (computer) graphical description. A project engineer will usually not discriminate lower levels than complete machines, and may even restrict considerations to plant. A materials scientist will usually discriminate down to molecules and atoms.

A set of other terms is always connected with the term "system", including purpose, behavior, structure, environment, input, output, property, and state.

SYSTEM - PURPOSE

Every artificial system serves a certain purpose. This purpose can be described with reference to a certain goal-system (a system of goals). If a *goal* is defined as a state of affairs that is thought to be possible, and that this state of affairs should preferably be realized, then a *goal system* is a set of goals and their relationships. Some of these goals may be hierarchically related to one another, in the form of "super-goals" and "sub-goals" (or partial goals). A sub-goal can concretize a super-goal, but more usually it already represents a means to realize that super-goal.

SYSTEM - BEHAVIOR

A *behavior* can be defined as a set of successively attained states of a system. For biological systems, behavior is interpreted as the sum of the reactions (to stimuli) of an organisms.

Behavior is not always a useful property of a system, for instance it is not applicable to systems of terms (terminologies) or systems of goals (goal-systems). For technical systems, achieving a certain behavior is generally the super-goal for creating the system.

The purposeful behavior of an object system is frequently termed its *function* (i.e. its teleological function [185]). Function then refers to something more stable, namely a desired ability to perform a set of actions (e.g. to be able to exert a set of effects). This desired ability may not be ensured by the actual behavior of the system, a system can also behave wrongly or badly. The term "behavior" can thus be applied in a broader sense than "function", the latter is always connected with the desired actions.

SYSTEM - STRUCTURE (Str)

The term "structure" is usually applied to the internal arrangement, order, organization, break-down, segmentation, conformation, constitution or construction of a system. In the same sense, one can speak of a network of elements.

Structure is therefore the set of elements in a system and the set of relationships that connect these elements to one another.

If $\quad El = (el_1, el_2, \ldots el_n)$
is the set of all elements,

and $\quad R = (r_1, r_2, \ldots r_n)$
is the set of all relationships,

then $\quad Str = (E, R)$
is the structure, the set consisting of **El** and **R**.

Structure and behavior (where applicable) are the most important properties of a system.

Many systems, and thus many structures, can be defined with respect to a single object. For example, the human body can be considered to contain a skeletal system (a structure of bones and their relationships), a circulatory system, a nervous system, an alimentary system, etc.

The word "structure" is usually used in one of three senses. The most comprehensive one refers to the term as used in this book, the set of elements and the set of their relationships. Particularly in philosophy, the same word is used to refer only to the set of relationships in a system. In engineering, the word "structure" is frequently used for the set of load-carrying static elements of a technical system that fulfill the partial functions of "connecting" and "supporting" - the frame, stand, housing, etc.

SYSTEM - RELATIONSHIP BETWEEN BEHAVIOR AND STRUCTURE
The behavior of a system is determined by its structure.

A relatively closed system with a given structure has only a single behavior; the structure determines that behavior.

Conversely, a behavior does not determine a unique structure. The same behavior (and function) can be achieved and realized by different structures.

SYSTEM - ENVIRONMENT (Env)
The *environment* of a system **S** is in theory everything that is not included in that system. Practically, considerations are restricted to the environment determined by the totality of all systems that conform to the following conditions:

- the system to be counted among the environment must have at least one element, and the output of that element must be simultaneously an input to an element of the system **S**, OR it must have at least one element that receives as its input an output from an element of the system **S**. This "close" or "direct" environment is termed the *active environment* of the system **S**.

Every system environment contains the following constituents: geosphere, atmosphere, biosphere (including humans), technosphere and astrosphere.

SYSTEM - INPUT (In), OUTPUT (Out)
The *input* represents the external relationship from the environment to the system. The input can be an effect, a coupling, or the measure of a property of a particular state of the operand, according to the type of system under consideration. The totality of all inputs can be collectively termed the total input, and represented by a vector (or tensor) of all individual inputs.

The *output* represents the external relationship from the system to the environment. The output can be an effect, a coupling, or the measure of a property of a particular state of the operand, according to the type of system under consideration. The totality of all outputs can be collectively termed the total output, and represented by a vector (or tensor) of all individual outputs. The output of a system is the set of those outputs of all elements that do not constitute inputs to other elements of that system.

The inputs and outputs (quantities) are the only connections between a system and its environment. It is emphasized that inputs and outputs represent all possible kinds of coupling: desired as well as undesired (disturbances, waste, etc.), and couplings that have the characters of material, energy or information (data or commands).

The meanings of the two terms "input" and "output" are not uniformly accepted. Some authors regard inputs and outputs as the respective elements of the system that will accept inputs (receptors) or deliver outputs (effectors). According to this interpretation, stimuli (the input effects of an environment, practically only information) enter a system through the inputs, and the reactions (answers, output effects of a system) emerge from the outputs as the system's effects on the environment. Other authors regard inputs/outputs as those things (i.e. practically only materials) taken in by the system from its environment, for instance nourishment for a biological system. The background for these differing opinions.may be seen in the aims and systems of images that have led to formulating the general ideas of systems.

SYSTEM - PROPERTIES (Pr)
SYSTEM - EVALUATION

Every system, and each of its elements and relationships, possesses a set of *properties.* These are an integral part of the system, and define it more precisely - size, mass, speed, form, stability - but also the less obvious properties such as suitability for manufacture and transport, and in particular the ability of the system to do something, its behavior.

No existing object is completely lacking in properties (that fact alone would be its property). Nevertheless, differences between systems will occur in the *state of embodiment* of a particular property, and its *measure* or numerical value. For instance, one of the properties of humans is labelled "height"; if a particular person is 170 cm tall, then the property "height" has measure 170 cm, the numerical value 170, with "cm" as the unit for that value. Another property is "having hair on head"; this property can have states of embodiment (according to the quality that is being measured) ranging from "none" to "much", or "blonde" to "black", and each of these qualities can be measured on a suitable scale to deliver a more or less precise measure (numerical value and unit of measurement). This discussion leads directly to the problems of quantifying some of the properties, see chapter 7.

A *property* is any characteristic or quality that is possessed by an arbitrary object, and that describes or characterizes that object.

This terminology needs further comment with respect to various criteria. Properties in the logic of predicates are treated by the unitary predicates. Ordinary language uses predicates to describe properties. In philosophy, the major properties are termed attributes. For technical products, the term *parameter* is used for those important properties that provide a closer description of the functions of that product. A parameter usually also provides information about numerical values, and how they change depending on the measures of other properties. A collection of such co-ordinated measures is in common engineering language termed a "characteristic", but this term is reserved for a somewhat broader meaning in the theory of technical systems, namely the collection of all such sets of coordinated measures of properties.

In order to fully characterize an object, for instance for the purpose of *evaluating* it, we search for characteristics that usually contain a number of properties and their measures. Measures that contribute to an evaluation are then termed *partial values*, which may be combined to a *total value* as the basis for judging the quality of that object. (The same words are used in philosophy and psychology, but with different meanings.) Individual properties must be quantified, and these partial values must be combined by suitable mathematical processes to obtain the total value.

SYSTEM - STATE

The total of the measures of all properties of a system at a given time is termed the *state* of the system.

The state of a system can be regarded as a vector (or tensor) which has the individual properties as its components, and is in this respect similar to a quality or total value. When determining the state of a system (or a measure of its quality), any non-significant or uninteresting properties are usually neglected.

Two systems (or one system in two states) can be either equivalent or not, any change in values between the two is termed the *difference*. A difference may exist either between two states of a system, or during the transition from one state to another. During such a transition, the difference can occur differentially (if the resulting transition is continuous) or in discrete steps.

SYSTEM - MODEL

The generalized model of a system shown in figure B.2 illustrates the above definitions and their relationships.

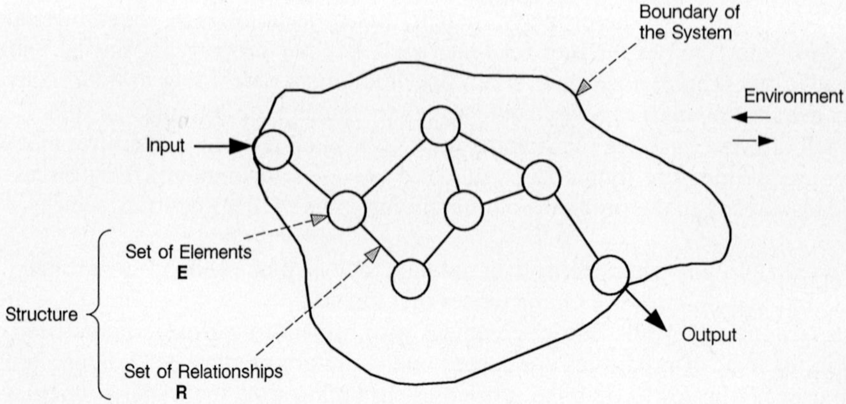

Fig. B.2 Model of a System

SYSTEM - TYPES

It is possible to discern many types of system according to various criteria. Some of the important classes of criteria are listed and defined here, for reference in the descriptive chapters of this book. Any one system can be classified in different ways by these criteria:

a) By the position of the system in the hierarchy of systems:
   - super-system, system, partial or sub-system.
b) By its connections to the environment:
   - open: with defined environment, and at least one input or output;
   - closed: with no connections to the environment.

(Note the differences between this definition and the one usually used in mathematical modelling of the dynamics of a system - in that definition an open-loop system has feedback by means of human operations, a closed-loop system has self-acting internal feedback.)

c) By the system's changes of state:
   - dynamic: its state is variable with time;
   - static: its state is constant in time (and this is a relative statement with respect to the observed properties).
d) By the system's anticipated behavior:
   - deterministic: a unique statement about the behavior of the system is possible, and depends only on the states of the system;
   - stochastic (probabilistic): only prediction of the variability of behavior is possible.
e) By the concreteness of the elements:
   - concrete: the elements are real objects or operations;
   - abstract: the elements are not real objects or operations.
f) By the origin of the system:
   - natural: occurring in nature;
   - artificial: produced and/or created by humans.
g) By the dependencies of its outputs:
   - combinatorial: the system's outputs depend only on its inputs;
   - sequential: the system's outputs depend on its inputs and on other quantities.
h) By the complexity of the system's structure:
   - extremely complex: for instance a brain, a country's economy;
   - very complex: for instance a fully automatic factory, a company's economy;
   - complex: for instance a passenger car, a city library;
   - simple: for instance a family library, a screw.
i) By the types of elements:
   - object (object system): elements are (abstract or real) things - for example house, motor, machine;
   - process (process system): elements are operations - for example manufacturing, classifying, destilling, cooking.

SYSTEM - TYPES OF TASK

Three types of task may be asked of the engineer with respect to systems:

Synthesis
- given: behavior and other requirements
- find: structure that satisfies the behavior and requirements

Analysis
- given: structure
- find: behavior

Black box problem

- given: the system, of which the structure is unknown or only partly known
- find: the behavior, and possibly the structure.

SYSTEM - SYMBOLIC REPRESENTATION

A system is represented on a diagram by geometric shapes consisting of rectangles, circular arcs, and combinations of these. For the purposes of this book, different symbols have been chosen for process systems and object systems, as shown in figure B.3 (see also the list of symbols).

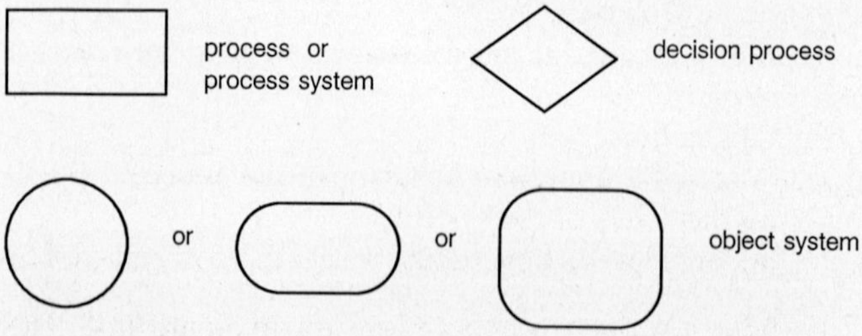

Fig. B.3 Graphical Symbols for Two Types of System

PROCESS (P)

Generally the term "process" refers to a procedure, a happening, a change in something during a period of time. In nature, some changes are always proceeding. Even on an object that appears to us to be very stable and unaltered (a mountain, a rock) suffers continual change by processes such as erosion, weathering, ageing, etc. We may also think of the evolutionary development of living matter, plants, animals, etc.

Beside these natural processes, humans organize various artificial processes to achieve the changes that are necessary or desirable. Such changes or transformations serve to satisfy human needs. The limitations imposed by the laws of nature must be respected, but some natural processes can be accelerated or retarded, and some properties can be improved, degraded, increased, or reduced.

The processes of producing purposeful changes of certain objects is essential for human life. These artificial processes are called transformations (consisting of operations), in which certain properties of an operand are subjected to a change. *Operand* (Od) is the general term used for all objects that are changed. The changes are accomplished according to the declared aims (goals), with the help of actions performed (and effects exerted) by humans, and usually also by technical systems, whereby the desired state of the operand is achieved.

A change in the operand (an operation) is the consequence of a certain effect, which is based on physical, chemical, biological and other phenomena (experiences). Each such effect is described by a rule, prescription, recipe, regulation, algorithm or technology. Typical bodies of knowledge about transformations in specific areas may be found in manufacturing, chemical processing, thermodynamics, etc.

A set of *operators* contribute to exerting the effects that are necessary for performing the transformations. The output of the operators are their effects on the operand. We use the general expression "effect", without distinguishing whether they are output effects (from an operator) or input effects (on an operand); action and reaction are equal and opposite.

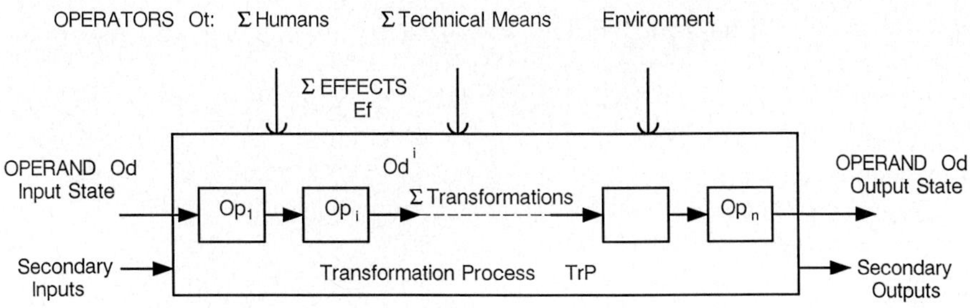

Fig. B.4 Model of a Transformation Process

The general model of a transformation process is shown in figure B.4. The effects of the operators consist of a flow of materials, energy and information (commands and data) that perform the changes on the operand. Further details about transformation systems is contained in chapter 3, and about processes in chapter 4.

This terminology is similar to the usage in mathematics. An operand (the input state) is transformed into a result (the output state after the transformation) by applying various operators, including "equals", "plus", "minus", "multiply by", "divide by", "differentiate", "sin", "cos", "tan", and many more.

If a process is considered to be a system of operations, then a uniquely determined sequence of operations is termed an *algorithm.* This sequence of operations may either be determined in advance once and for all times, or it may depend on the results of a preceding operation (known for instance as adaptive control). The second alternative demands that certain conditions from the preceding operations are fulfilled before the next operation can be started. Thus an algorithm can be defined as a set of operations and their relationships, namely as the structure of the process. This is similar to the term "technology", which can be regarded as a collection (or set) of the individual working operations needed to manufacture a certain workpiece or material.

"Regulating" and "controlling" are two important and typical members of the many existing *types* of process: *Regulating* is a process in a system whereby one or more magnitudes (as input values) influence a magnitude (as output value) by means of the governing rules of that system. It is characteristic for an open time-sequence of events, the regulating chain. *Controlling* is a process in a system whereby a magnitude (the output magnitude to be controlled) is continuously monitored or sampled, and compared to another magnitude (a desired magnitude or a standard), and depending on the result of that comparison the input magnitude is influenced with the aim of causing a subsequent comparison of the output values to be more favorable. The resulting sequence of events takes place in a control loop which is closed either by a human operator (open-loop control), or by a technical system (self-acting, closed-loop or automatic control).

RELATIONSHIP (R)

A relationship is the real or meaningful dependence or interaction between two or more objects or phenomena of abstract or concrete kind. Relationships connect individual elements into a system, different relationships form different systems. Objective, exact and describable relationships of the natural sciences are important for designing.

RELATIONSHIP - PROPERTIES

The statement "object **X** is in a relationship to object **Y**" is expressed in symbols as **R (X, Y)**. Such relationships can be reflexive, symmetrical or transitive.

- Reflexivity: every object is equivalent to itself.
- Symmetry: if one object is equivalent to a second one, then the second is also equivalent to the first.
- Transitivity: if two objects are each equivalent to a third, then they are also equivalent to each other.

If all three properties are fulfilled, the relationship is termed "equivalence". The relationship between two objects is also termed "correlation", which is a mathematical representation of the most general form of connection.

RELATIONSHIP - TYPES

a) **Similarity** - a characteristic and expressed relationship between two or more systems (objects, processes, statements) based on certain properties that they have in common. It ranges from complete equality (identity, equivalence) to a very remote similarity. Similarity can concern function, structure, form, etc. This is therefore an extended interpretation from that of "equivalence of form (shape)" without equivalence of magnitude (size).

Similarity is important for modelling and experiments on models. Laws of similarity should formulate the conditions under which experiments on models are valid with respect to the modelled reality. For example, aerodynamic flows are similar if Reynold's numbers are equal. The conditions of similarity then deliver concrete conditions under which the results of an experiment can be transferred from a model to the full-sized reality.

The range of similarity can be defined as the intersection of the sets of properties of systems that are in a relationship to each other.

b) **Analogy** - co-incidence of objects of phenomena in important characteristics and properties, particularly in the functions and structures of systems. It is a form of similarity, and the term is often used in this sense.

c) **Homomorphy** - relationship between two systems in which every component (or element) and every relationship in one system permits conclusions about the elements and relationships in the second system, but conversely conclusions about the first system cannot be derived from the second.

The comments about conditions for similarity to "scale up" the results of experiments on models are also valid here, such models are homomorphic with the large-scale systems being modelled.

d) **Isomorphy** - relationship between two systems in which every component (or element) and every relationship in one system can be uniquely mapped to a component (or element) or relationship of the other system, and the converse mapping also holds (a symmetry condition).
e) **Equivalence** of object - objects are termed equivalent if an equivalence relationship exists between them. Similarly, equivalence can refer to sets or statements (as the relevant objects). This relationship is also understood as equality of value, including in monetary terms (e.g. using a commodity as a value standard to designate amounts of another commodity).
f) **Identity** - relationship between objects, states, statements, etc. in which there is complete coincidence of some characteristics between them. Absolute identity demands coincidence in all characteristics. Relative identity demands coincidence in only some characteristics, and is usually termed "similarity".

Strictly speaking, identity refers only to the property of reflexivity, whereas equivalence also implies symmetry and transitivity. In engineering, both terms are usually used as synonyms.

g) **Mathematical functions** - these are an important class of relationships that express a regular inter-dependence among variables:

$y = f(x)$ ... equality, equation
$y > f(x)$ ... inequality, limiting
also $<$, $\neq$, etc.

A mathematical function expresses an exact relationship between the variables **x** and **y**, a deterministic connection.
h) **Causality, causal chains** - a two-valued assymetric relationship between a cause and its consequence. The cause always precedes the consequence. Causal relationships can appear in strict (deterministic, if-then, or unconditional, iff-then, i.e. "if-and-only-if"-then) or in various weakened forms. Causal chains occur when a consequence of a previous causal event appears as the cause of a further event.
i) **Coupling** - occurs when certain outputs of an element (or system) are simultaneously inputs to another element (or system).

Coupling can be in series, in parallel, or as feedback, or in any combination of these, see figure B.5. Coupling can be with respect to materials, energy or information (commands or data).

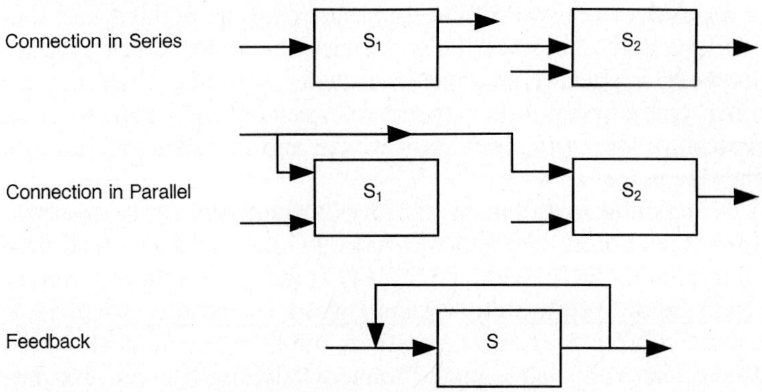

Fig. B.5 Types of Coupling Between Systems

j) **Goal-means relationship** – a two-valued assymetric relationship between a system of goals (aims, purposes, tasks) and the system of means by which the goals can be realized.
k) **Spatial relationship** – describes the mutual location in space, and therefore the arrangement of elements that participate in the relationship. The branch of knowledge concerned with spatial relationships *per se* is topology. For practical application, this knowledge is concretized by geometry, geography and other branches of knowledge.
l) **Logical relationship** – in general, these designate relationships between objects (in the study of logic as two- or multi-valued predicates), for instance:

e_1 is smaller than e_2
e_3 is beside e_4

The best-known constants (functions) are: and, or, and/or, not, true, not true, therefore, either/or, if - then, only if - then, exactly if - then, is equal. This list shows that some of the previously defined relationships are also logical relationships.

These relationships are realized in computers with the help of logical elements, in particular using Boolean algebra, which represents a formal treatment of logical relationships.
m) **Time relationship** – describes the arrangement and progression of processes, occurrences, happenings, etc. along the axis of elapsing time.

B.3 Choice of Words

Whenever a novel idea, technique, theory, etc. is developed, the authors have a choice of (a) using established words from existing terminology in the new context, or (b) selecting or coining new words to better express the thoughts and concepts.

This book is no exception. The theory of technical systems describes many items of existing technology (in the wider sense of that word, see also chapter 1). We have chosen, at times, to adopt established words, but to define them in a more precise and restrictive way (within the usual meanings as defined in a dictionary) to suit our purpose. At other times we have adopted words that are unusual in the context, especially to convey a wider meaning than usually attached to the word. This section of the appendix is intended to explain some of the reasons for choosing some of the important words, and to make comparisons to usages in another language.

The entries are by intention in alphabetical order (but not very strict), this presented the least problems. Where German words are substantially different, they are given in italics (e.g. SCIENCE *WISSENSCHAFT*), but where there is only a relatively minor change of spelling, the German word is not mentioned (e.g. STRUCTURE *STRUKTUR*). Previous expressions published in English books and papers by WDK authors are also quoted immediately after the current preferred term.

In this book, we have normally tried to avoid stringing two or more nouns together (with the first implied as adjectives) so that we can avoid any ambiguity that results from multiple nouns. The text may therefore appear to labor in places, but if clarity is maintained or enhanced we make no apologies. The particular combinations of nouns listed here are deliberately exceptions to that self-imposed rule, they are intended as composite nouns, and could have been written with hyphens.

We have also tried to give structure (words as elements, their position in the sentence as the relationships) to the sentences such that any activity is referred to by a verb (e.g. designing). The sentence construction "the design of something" is usually replaced by (the activity of) "designing something", unless other considerations prevail.

ACTION MEDIUM - medium capable of transmitting an action. It is not necessarily an "active medium", it may at times be inactive, passive, or countering the desired actions.

ACTION CHAIN *WIRKKETTE* - chain of elements capable of performing an action.

ACTION PROCESS *WIRKPROZESS* - process that is capable of resulting in an action.

ACTION SITE, Action Locality, Action Location *WIRKSTELLE* - point, line, surface, volume, etc., at which an action may take place, namely the output of a TS where it exerts its effects on an operand. The word "site" was selected for this application because it coincides best with the German, for instance "Baustelle" is "building site". Previous publications (e.g. [119]) contained "locality" or "location", we now think "site" is better.

CHARACTERISTICS, PROPERTIES *EIGENSCHAFTEN* - Characteristics provide general statements about the things that characterize any object of investigation. Properties are the states of embodiment of these characteristics, and are therefore more specific, frequently with measures (numerical values) as their states of embodiment.

COMPLICATION, COMPLEXITY - these are usually used as interchangeable terms. We have chosen to imply a substantial difference. "Complexity" is used to denote that the system consists of many components. "Complication" can be a property of a single component, implying that it is not simple, and is probably difficult to produce. Complex systems may result from combining many simple components, or may incorporate some complicated components.

COMPONENT STRUCTURE, Anatomic Structure, Constructional structure *BAUSTRUKTUR, AUFBAUSTRUKTUR* - structure consisting of components and their relationships. In analogy with biological practice, previous publications (particularly [119]) used "anatomy" to designate the equivalents of bones, muscles, etc., the collection of recognizable physical elements. Other people have tended to use word combinations which seem clumsy, e.g. "constructional structure", the structure of constructional elements, the structure which is constructed by making and assembling the items of hardware. The compromise reached for this book is "component structure", consisting of "components" (sometimes also termed "constructional elements") and their relationships. The alternative word for component is more applicable when such an element can be conceptually recognized in a structure, but it is difficult physically to separate it from its connected neighbors – can a column or a beam be taken out of a civil engineering structure made of reinforced concrete?

CONCEPT – This word is used in two meanings. One is in the spirit of the dictionary definition as a general idea about a class of objects, and refers to mental pictures and maps. The other meaning is more specific in terms of engineering design, where a concept is defined as a line-diagram of a proposal for solving a problem. The organs are visible, but not necessarily in their final (geometric) arrangement, and few (if any) real dimensions are specified. This form of concept is a transition stage from a function structure (possibly by way of an organ structure) towards a preliminary layout of a TS.

CONCRETIZE – the process of making an idea or concept more concrete, working towards realizing it. This activity takes the opposite direction to that of abstracting.

CONSEQUENCE *WIRKUNG* – "cause and effect" are the words usually linked together in normal language. We have reserved the word "effect" for the acting output of a technical system, and have therefore altered the common expression to "cause and consequence". The German "Wirkung" is closer to the English "action", and is also used in "Wirkbewegung" (action motion), "Wirkweise" (mode of action), "Wirklich" (real), etc. The dilemma is obvious.

DESIGN *KONSTRUKTION* – may refer (a) to the results (appearance, repetitive pattern, etc.), or (b) to the activity of creating and specifying the results. In this book, the word is avoided where possible because of this potential ambiguity (does it refer to the resulting pattern, or to the activity of newly producing it?).

DESIGN FOR FUNCTION, DESIGN FOR MANUFACTURE, DESIGN FOR CORROSION, ETC. *FUNKTIONSGERECHTES KONSTRUIEREN, etc.* – Rules and guidelines (heuristics) for designers, so that the TS to be designed (a) will be suitable to perform the necessary functions (and therefore exert the needed effects) to drive the TP, (b) will be economical and easy to manufacture, (c) will resist corrosion, etc. The German language has an easier expression, "Funktionsgerecht" literally translates to "just, right, fair, fit for function".

DEVELOPMENT, DEVELOPMENTS *ENTWICKLUNG* – One of the processes following that of designing a technical system, and manufacturing a prototype of it, is development (singular), defined as testing and developing a TS to bring it up to the required performance and other properties. Successive families of TS show changes from one model to the following one, which demonstrate the developments (plural) that have taken place in the state of the art and other measures of technical progress. We could have used "advancements", but some of these developments seem to be steps backwards.

EFFECT – the word "effort" usually implies force. We intended a wider interpretation that can include many other kinds of "effort" as the acting output that a technical system delivers to the operand of a process. These effects can contain or consist of material, energy and/or information. The location from which a technical system delivers an effect is therefore the EFFECTOR, similarly the TS receives its inputs at RECEPTOR locations.

ESTABLISH *BESTIMMEN* – the usual action is to "determine" something, which implies a more analytical frame of mind and a single valid result. In designing, many choices are open, and any one choice has a range of validity and appropriateness depending on the circumstances of the choices to be made, and the person doing the choosing. This is not a "deterministic" process, we have therefore chosen to use the term "to establish" to describe the process of eliminating (or deferring from consideration) those solutions that are not considered appropriate or optimal. The German implies making something definite.

FORM *FORM, GESTALT* – for purely mechanical engineering purposes, the term "shape" is more usual when applied to the form of a solid object, and refers to its macro-geometry. The micro-geometry and other properties are usually not considered in the term "shape" – tolerance, surface roughness and waviness, deviations from nominal geometry, internal structure, etc. Operands in technical processes, and inputs/outputs of technical systems, may exist as materials, energy and/or information. Of these, only materials (in a solid state) can have "shape", but all three can exist in different forms. The word "form" is therefore adopted as the more general term. The German "Gestalt" contains some additional meaning as stature, figure or build, and implies also dimension and size. Derived from that word is the verb "gestalten", to give form to something, including its arrangement, organization, configuration. This term can also be used to describe the process of giving form to information. No good English equivalent exists, but poor seconds are available in "form-determination", "embodiment", etc.

FUNCTION – may refer (a) to the way some technical system works to do something, or (b) to what it does. In this book, we have distinguished the internal actions (how it works, its **mode of action**, resulting from its action chains and action processes) from its external actions (what it does or is intended to do, its **function**, its exerted effects resulting from the externally established functions it is intended to perform).

FUNCTION CARRIER *FUNKTIONSTRÄGER* – organ that carries specific functions.

FUNCTION STRUCTURE, Functional Structure – structure consisting of functions and their relationships, structure of functioning, structure of activities. Previous publications have used "functional structure", but objections have been raised because a proposed structure of this kind may not actually be able to work, it may be "non-functional". We have therefore chosen to use the words "function structure", a structure consisting of functions (as its elements) and their relationships.

GOALS, AIMS *ZIEL* – The German word "Ziel" translates to aim, end, target, object (or rather objective), mark, destination, finish, goal, purpose. Of these

English words, we think that "goals" or "aims" best expresses our intended meaning, and are used interchangeably. "Goals" is usually used in the combination "goal-system", "aims" is usually combined with "means" - it seems to sound better that way.

HEURISTIC - any knowledge or method that is useful towards a goal. A narrower usage exists, especially in European scientific circles, that restricts the term to a study (implying also teachability) of finding empirical solutions by established rules of thumb. It results in the concept of an heuristic algorithm, one that consists of an open chain of events, analogous to an unsecured (unsure) hypothesis.

LOCATION *ORT* - The German "Ort" is used for place, village, township, site, spot, locality. Of these, the most general and useful word seemed to be "location".

MEANS *MITTEL* - something by which an end (goal or aim) is achieved. We have tried to always use the plural, and avoid this word in its other meaning (whatever it may mean).

MEASURE, VALUE *MASS, WERT* - most measurable things are measured against a scale with a unit, the resulting number is a "measure" of the property (similar to the German "Maß"). Value tends to imply money, just as the German "Wert". "Magnitude" implies something not very accurate, especially when it is an order.

METHOD, METHODOLOGY - The second word is frequently mis-used to name an entity of the first kind. A method may be a very complicated sequence of recommended steps in a procedure or mental activity. A methodology only exists when various methods are brought into a logical connection and given a scientific basis. We make a strict distinction between these concepts.

MOTOR, ENGINE - different branches of engineering use different terms to denote a "source of appropriate mechanical energy", a "prime mover". In motor vehicles, the term usually used is "engine", an ironic transformation of terms within a branch of engineering. Other branches of engineering use the term "engine" as a product of ingenuity - a ruling engine scribes regular lines on a measuring bar. In this book, all "sources of energy to produce motion" for the immediate application are termed "motor", without regard to the nature of the input energy that they convert to (rotational or translational) mechanical energy.

OBJECT SYSTEM *SACHSYSTEM* - a system consisting of abstract or real objects (hardware), as the object to be designed, considered or handled. It will accept inputs, and produces effects at its action sites, and other outputs at some connections to the environment. A technical system is a special form of object system, one produced by human effort. The German "Sache" translates to thing, object, affair, matter, business, concern, circumstance, fact, point, issue, case.

OPERATION - We are concerned with three (somewhat different) interpretations, (a) an operation - a small part of a technical process, (b) in operation - an active state of a technical system, and (c) being operated - acted on by members of the human system to cause a TS to enter an active state, or control it while in that state, and thus drive an operation in the TP.

ORGAN STRUCTURE - structure consisting of organs and their relationships.

PARTIAL TRANSFORMATION, PARTIAL OPERATION, PARTIAL FUNCTION, PARTIAL SYSTEM, PARTIAL PROCESS, PARTIAL ORGAN *TEIL-TRANSFORMATION, etc.* - a transformation, operation, function, system, process or organ that constitutes a part of a transformation ... organ at a more complex level. Such a partial transformation ... organ may or may not be able to exist by itself. It may reach the simplest possible level, an elementary or elemental transformation ... organ. We have used a consistent way of expressing these subdivisions, using the adjectives "partial" and (for the limit) elementary or elemental. It may be argued that some of the subjects of these terms (e.g. function) cannot be sub-divided into something that is "partial", or that they represent something at a lower (inferior) level of interest. The term "subsystem" is frequently used in this context, but the prefix "sub-" does not readily provide a consistent notation. We intend to imply that each of these concepts may be sub-divided into contributing parts. These subdivisions may in turn be the primary concern of other engineers, who may wish to sub-divide for their own purposes. The German "Teil" means "part of", a sub-division, as well as "component".

REALIZE - the process of making an idea or concept into a working reality.

REGULATE, CONTROL *STEUERN, REGELN* - We have tried to make a distinction between the executive function of input to achieve an initial direction, regulating, and measuring the output to correct it by adjustments fed to the input, controlling. Firing a brand-new sports rifle for the first time at a target requires regulating, there is no previous reference to be able to predict the properties of the system and its actual performance. Firing the second shot to achieve a better score, and adjusting the sights, are already control actions. Regulating is usually subsumed into controlling. The German "steuern" means "to steer", and is distinct from "regeln" and "kontrollieren", the last of these means "to check up on".

SCIENCE *WISSENSCHAFT, LEHRE* - the assumptions of science usually implied in English language usage are that a subject must be capable of corroboration by observation, experiment and inference. They are rendered more severe by also requiring that the results can be verified by independent experiment, and mathematical models. In this book, we adopt the wider implications of the German term "Wissenschaft" (body of knowledge), which includes all categorized and related knowledge. Science does not appear to be limited by the subjects that can (or can not) be studied, but rather by the mental procedures involved in that study. A particular form of science is the knowledge adapted to *teaching*, which in German is known as "Lehre" (e.g. Festigkeitslehre = knowledge of mechanics of materials adapted to teaching).

STANDARD, STANDARDIZED *NORM* - Many components, measures, values or procedures are defined in accepted standards, documents issued by recognized corporate authorities - these components are labelled "standard" in this book. Other components are selected from an available wider range for the internal purposes of a smaller organization that does not constitute a recognized authority - these are labelled "standardized". The distinction may or may not be useful in a certain context, it is made here to show the power of authority behind such a selection.

STATE, CONDITION *ZUSTAND, BEDINGUNG* - The word "condition" has an overtone of "conditional", something on which a state of affairs depends. These two words are strictly separated in this book. "State" refers to the way that a TS or TP exists at any one time, including its properties and their measures. "Condition" is used to make statements about imposed circumstances that are essential to the existence of a TS or TP, especially limits. "Constraints" are a form of condition. The analogy with the German is fully intended.

STATE OF EMBODIMENT - Most characteristics and properties can exist in various measures, values and states. A "color", as the general concept, can exist as "blue", "red", etc. as well as in various intensities. The German word "Ausprägung" literally translates into "stamped out, coined, formed, impressed". We have chosen the expression "state of embodiment" to name the possible discrete ways in which the general quality can appear in an embodied form - especially because the word "form" has already been reserved for a specific use.

STRUCTURE - the union of the set of elements (which may be conceptual or real) and the set of their relationships. In the English language of engineering, the word "structure" is frequently used for the set of load-carrying static elements of a technical system that fulfill the partial functions of "connecting" and "supporting" - the frame, stand, housing, etc. This second usage of the word is avoided.

SYSTEM - a set of connected parts, with the relationships that connect them (internally), and the connectivity to the surroundings. Various structures may be recognizable, by selecting a few characteristics or properties and showing the nature of the connecting relationships, and the regularities followed by them.

TECHNOLOGY - as a generic term (without a definite or indefinite article) the usual interpretation is similar to the German word "Technik", and is increasingly used to denote any application of knowledge to enable designing and producing technical systems and processes. In most cases we have adopted a direct translation of the German word "Technologie", with a definite or indefinite article. Thus a (or the) technology is the specific way of delivering an effect to an operand.

TELEOLOGY, CAUSALITY, FINALITY - The arguments about these concepts are largely philosophical, and concern mainly the extremes of interpretation. If we cannot rely on a set of cause-consequence chains, we can do nothing to design or control our environment. Causality is thus an essential assumption to engineering. Finality is equally important for engineering, especially to explain engineering design. It is not necessary that ALL events (as consequences) MUST have IDENTIFIABLE causes, just those that we wish to influence in a predictable way. With this book, we intend to influence some important aspects of engineering, particularly the processes of designing and teaching/learning. For this purpose, we suggest that a teleological approach to those aspects that can be influenced is necessary, and that therefore the other two concepts are needed by engineers.

TP PHASES - An earlier book in the English language [119] used some expressions that we no longer consider adequate. In that work, the three essential phases of a technical process were defined as "preparation", "execution" and "conclusion". We now think it better to define these phases by what happens, "preparing", "executing" and "finishing" operations as essential parts of a complete transformation. The name used for the final phase in the earlier work seemed to imply that conclusions can be drawn, just as in research work, which is too restrictive for our purposes.

Bibliography

1. –, Academic American Encyclopedia, **8** "F", p. 230–232, & **12** "K-L", p. 394–395, Princeton, N.J.: Aretê Publ., 1981
2. –, The Canadian Encyclopedia, p. 664–673, Edmonton: Hurtig Publ., 1985
3. –, The Concise Oxford Dictionary, (7 ed), Oxford: Univ. Press, 1982
4. –, Encyclopedia Americana, **17** "Latin America – Lytton", p. 844–849, Danbury, Conn.: Grolier, 1983
5. –, Encyclopaedia Britannica, The New (15 ed), Micropedia **V** "Hermoup-Lally", p. 22, & Macropedia **18** "Taylor-Utah", p. 352–358 (esp. p. 354), Chicago: Benton, 1974
6. –, ESDU Index (published annually), Engineering Sciences Data Unit, London, England
7. –, McGraw-Hill Encyclopedia of Science and Technology (5 ed), **7** "ice-lyt", p. 795–798, New York: McGraw-Hill, 1982
8. –, The Measurement of Scientific and Technical Activities, 'Frascati Manual', OECD, 1981
9. –, The Merriam-Webster Dictionary, New York: Pocket Books, 1974
10. –, Normen Auszug für Technische Schulen (8 ed), Zürich: VSM Normenbüro, 1978
11. –, The Practice of and Education for Engineering Design Conference, Proc. I. Mech. E. **178** (1964) Pt. 3 B, p. 34–35 & 51–52
12. –, VDI Richtlinie 2220: Produktplanung – Ablauf, Begriffe und Organisation, Düsseldorf: VDI, 1980
13. –, VDI Guideline 2221: Systematic Approach to the Design of Technical Systems and Products, Düsseldorf: VDI, 1987 (edited by K. M. Wallace)
14. –, VDI Richtlinie 2222 (1): Konstruktionsmethodik: Konzipieren technischer Produkte, Düsseldorf: VDI, 1977
15. –, VDI Richtlinie 2222 (2): Konstruktionsmethodik: Erstellung und Anwendung von Konstruktionskatalogen, Düsseldorf: VDI, 1982
16. –, VDI Richtlinie 2223: Begriffe und Bezeichnungen im Konstruktionsbereich, Düsseldorf: VDI, 1969
17. –, VDI Richtlinie 2224: Formgebung technischer Erzeugnisse, Düsseldorf: VDI, 1960
18. –, VDI Richtlinie 2225 (1) & (2): Technisch-wirtschaftliches Konstruieren, Düsseldorf: VDI, 1977
19. –, VDI Richtlinie 2234: Wirtschaftliche Grundlagen für den Konstrukteur, Düsseldorf: VDI, 1982
20. –, VDI Richtlinie 2235: Wirtschaftliche Entscheidungen beim Konstruieren, Methoden und Hilfen, Düsseldorf: VDI, 1982
21. –, VDI Richtlinie 2242 (1) & (2): Ergonomiegerechtes Konstruieren, Düsseldorf: VDI, 1983
22. –, VDI Richtlinie 2802: Kostenvergleichsrechnung bei Wertanalysen, Düsseldorf: VDI, 1976

23. -, VDI Richtlinie 3258: Kostenrechnung mit Maschinenstundensätzen: Begriffe, Bezeichnungen, Zusammenhänge, Düsseldorf: VDI, 1962
24. Adams, J.L., Conceptual Blockbusting (2 ed), San Francisco: Freeman, 1980
25. Alexander, E.R., "The Design of Alternatives in Organizational Contexts: A Pilot Study", Admin. Sci. Quart. **24** (1979), p.382-404
26. Alexander, E.R., "Design in the Decision-Making Process", Policy Sciences **14** (1982), p.279-292
27. Andreasen, M.M., "Darstellungsmöglichkeiten beim Konzipieren", Schw. Masch. **78** (1978) No.4 & 6
28. Andreasen, M.M., Synthesemethoder på Systemgrundlag - Bidrag til en Konstruktionsteori, Lund: T.H. (Doctor Diss.), 1980
29. Andreasen, M.M., & Hubka, V., WDK 4 a: Methodisches Konstruieren von Maschinensystemen - Fallbeispiele, Zürich: Heurista, 1981 (English edition in progress, see also [121, 122])
30. Andreasen, M.M. (ed), Workshop ATTS - Allgemeine Theorie Technischer Systeme (WDK Arbeitsdokumente), Lyngby: DTH-AMT, 1982
31. Andreasen, M.M., "Konstruktionsmerkmale und ihre Ausprägung", in [120], p.420-428
32. Andreasen, M.M., Kähler, S., & Lund, T., Design for Assembly, London: IFS Publ., & Berlin/Heidelberg: Springer, 1983
33. Artobolevskii, I.I., "Past, Present and Future of the Theory of Machines and Mechanisms", Mech. & Mach. Th. **11** (1976) Nr.6, p. 353-61
34. Artobolevskii, I.I., "General Problems in the Theory of Machines and Mechanisms", Mech. & Mach. Th. **10** (1975), Nr.2/3
35. Ashby, W.R., An Introduction to Cybernetics, London: Methuen Univ. Paperbacks, 1968
36. Ashford, F., The Aesthetics of Engineering Design, London: Business Books, 1969
37. Asimow, M., Introduction to Design, Englewood Cliffs, N.J.: Prentice Hall, 1962
38. Baynes, K., & Pugh, F., The Art of the Engineer, Guildford: Lutterworth, 1981
39. Beer, S., Cybernetics and Management, New York: Wiley, 1966
40. Beitz, W., "Grundlagen Methodischen Vorgehens beim Konstruieren", in [116], p.115-130
41. Beyer, W.H. (ed), CRC Handbook of Tables for Probability and Statistics (2 ed), Cleveland OH: Chemical Rubber Co., 1968
42. Birkhoff, G., Hydrodynamics: A Study in Logic, Fact and Similitude (2 ed), Princeton, NJ: Princeton Univ. Press, 1960
43. Bjørke Øyvind, Computer-Aided Tolerancing, Trondheim: Tapir, 1978
44. Booker, P.J., "Principles and Precedents in Engineering Design", London: Inst. of Engineering Designers, 1962
45. Booker, P.J., History of Engineering Drawing, London: Northgate, 1979
46. Buck, C.H., Problems of Product Design and Development, Oxford: Pergamon, 1963
47. Buffa, E.S., Modern Production Management (3 ed), New York: Wiley, 1969
48. Burgess, J.A., Design Assurance for Engineers and Managers, New York: Marcel Dekker, 1984
49. Burington, R.S., & May, D.C., Handbook of Probability and Statistics with Tables (2 ed), New York: McGraw-Hill, 1970
50. Cox, H.L., The Design of Structures of Least Weight, Oxford: Pergamon, 1965
51. Davis, L.E., & Taylor, J.C. (eds), Design of Jobs, Harmondsworth: Penguin, 1972
52. DeBono, E., Lateral Thinking, Harmondsworth: Penguin, 1977
53. DeBono, E., Tactics: The Art and Science of Success, London: Collins, 1985
54. DeGreene, K.B., Sociotechnical Systems, Englewood Cliffs, NJ: Prentice Hall, 1973
55. Drucker, P.F., The Effective Executive, New York: Harper & Row, 1964
56. Dym, C.L., "Expert Systems: New Approaches to Computer-aided Engineering", Engineering with Computers **1** (1985) No.1, p.9-25
57. Earle, J.H., Engineering Design Graphics (3 ed), Reading MA: Addison Wesley, 1977
58. Eder, W.E., & Gosling, W., Mechanical Systems Design, Commonwealth and International Library 308, Oxford: Pergamon Press, 1965
59. Eder, W.E., "Design Engineering Methodologies: Comparisons of English- and German-Language Regions and Influences of Culture", in [116], p. 41-53
60. Eder, W.E., "A Viewpoint on the Quantity 'Plane Angle'", Metrologia **18**, (1982) p.1-12 (plus Errata, Metrologia **18**, (1982) p.171)

61. Eder, W.E., "A Viewpoint on the Quantity 'Solid Angle'", Metrologia 19 No. 1, 1983, p. 1-8
62. Eder, W.E., "Design Education: Survey and Problems", in [124], p. 989-998 a
63. Eder, W.E., "Engineering Design as a Curriculum Organiser", in L.P. Grayson, & J.M. Biedenbach (eds), Partnership with Industry and Community, Proc. ASEE Annual Conference 1986, Washington: ASEE, 1986, p. 629-637
64. Eder, W.E., "Engineering Design - Practice and Theory", New York: ASME, Paper 86-DET-113, Design Engineering Technical Conference, October 1986
65. Eder, W.E., "Structures as Models in the Design and Development of a System", in H. Yoshikawa & E.A. Warman (eds), Design Theory for CAD, Proc. IFIP W.G. 5.2, Amsterdam: North Holland, 1987, p. 33-50
66. Eder, W.E., "Learning Reality from Examples: Case Studies Compared", Int. J. Appl. Engng. Ed. **3**, No. 1, (1987) p. 5-9
67. Eder, W.E., "Design Science - A Survey of Some Approaches", in L.P. Grayson & J.M. Biedenbach (eds), Engineering Focuses on Excellence, Proc. ASEE Annual Conference, Washington, DC: A.S.E.E., 1987, p. 668-674
68. Eder, W.E. (ed), WDK 13: Proceedings of the 1987 International Conference on Engineering Design, ICED 87 Boston, New York: ASME, 1987 (2 vols)
69. Eder, W.E., "Theory of Technical Systems - Prerequisite to Design Theory", in [68], p. 103-113
70. Eder, W.E., "Survey of Design Education", in [68], p. 881-887
71. Edholm, O.G., The Biology of Work, New York: McGraw-Hill World University Library, 1967
72. Ehrlenspiel, K., "Möglichkeit zum Senken der Produktkosten - Erkenntnis aus einer Auswertung von Wertanalysen", Konstruktion **32,** (1980) Nr. 5, p. 173-178
73. Ehrlenspiel, K., Kostengünstiges Konstruieren, Berlin/Heidelberg: Springer, 1985
74. Ehrlenspiel, K., & Lindemann, U., "Ein Beitrag zur Theorie des Konstruktionsprozesses", Konstruktion **31,** (1981) Nr. 7
75. Eversheim, W., "Analytische Betrachtung von Konstruktionsaufgaben", Industrie-Anzeiger 91 (1969), Nr. 87
76. Farrar, D.J., "Britain's Unsung Wealth Creator", CME, March 1986, p. 24-26
77. Feilden, G.B.R., "Engineering Design", Report of Royal Commission, London: HMSO, 1963
78. Fendler, E., "Über gleichartige Änderungen der Wachsfunktion in der Entwicklung des Lebens und der Technik", VDI-Z **112** (1970), Nr. 2
79. Ferreirinha, P., "Herstellkostenberechnung von Maschinenteilen in der Entwurfsphase mit dem HKB Programm", in [124], p. 461-467
80. Fox, R.L., Optimization Methods for Engineering Design, Reading MA: Addison-Wesley, 1971
81. French M.J., Conceptual Design for Engineers (Revised edition), London: Design Council, & Berlin/Heidelberg: Springer, 1985
82. French, T.E., & Vierck, C.J., Engineering Drawing (10 ed), New York: McGraw-Hill, 1960
83. Frost, M., Value for Money - the techniques of cost-benefit analysis, Aldershot: Gower, 1971
84. Gage, W.L., Value Analysis, London: McGraw-Hill, 1967
85. Galletly, G.D., "Design equations for preventing buckling in fabricated torispherical shells subject to internal pressure", London: Proc. I. Mech. E. **200** (1986) No A2, p. 127-139
86. Gasparski, W., Understanding Design: The Praxiological-Systemic Perspective, Seaside, CA: Intersystems Publ., 1984
87. Giesecke, E.F., Mitchell, A., Spencer, H.C., Hill, I.L., & Loving, R.O., Engineering Graphics, New York: Macmillan, 1969
88. Gomez. P., & Probst, G.J.B., "Organizational Closure in Management: I. Complimentary View to Contingency Approaches", International Cybernetics Newsletter, Winter 1985, p. 703-710 (in Cybernetics and Systems, **16** (1985) No. 1)
89. Gordon, W.J.J., Synectics: the development of creative capacity, New York: Collier, 1961
90. Gosling, W., The Design of Engineering Systems, London: Heywood, 1962
91. Green, A.E., & Bourne, A.J., Reliability Technology, London: Wiley, 1972
92. Gregory, S.A., 'Design Science', in [93], Chap. 35, p. 323-330
93. Gregory, S.A. (ed), The Design Method, London: Butterworths, 1966

94. Greve, W., Wertanalyse in der Fertigung, Stuttgart: Berliner Union, 1969
95. Habacher, M., Österreichische Erfinder, Wien: Bergland, 1964
96. Hajek, V.G., Project Engineering (Profitable Technical Program Management), New York: McGraw-Hill, 1965
97. Hald, A., Statistical Theory with Engineering Applications, New York: Wiley, 1952
98. Hall, A.D., A Methodology for Systems Engineering, Princeton, N.J.: Van Nostrand, 1962
99. Hansen F., Konstruktionssystematik (2 ed), Berlin: V.E.B. Verl. Technik, 1966
100. Hansen, F., Konstruktionswissenschaft - Grundlagen und Methoden, München: Carl Hanser, 1974
101. Hansen, F., Adjustment of Precision Mechanisms, London: Iliffe, 1970
102. Harrisberger, L., Engineersmanship: a philosophy of design, Belmont, Calif.: Brooks/Cole, 1966
103. Hertzberg, F., Work and the Nature of Man, New York: Harcourt Brace & World, 1966
104. Hertzberg, F., "One more time: how do you motivate employees?", Harvard Business Review **46** (1968) p.53-62
105. Hertzberg, F., Mausner, B., & Snyderman, B., The Motivation to Work, New York: Wiley, 1959
106. Homann, H.W., Gührling, H., & Brankamp, N., "Ein Klassifizierungssystem für Stahlbaueinzelteile - Entwicklung und Beschreibung des Systems", Industrie-Anzeiger **20** (1970), Nr.6
107. Howard, G., Evolution of Motorcar Shapes and Design, Genève: Inderscience, 1985
108. Hubka, V., "Grundlegender Algorithmus für die Lösung von Konstruktionsaufgaben", in XII Internationales wissenschaftliches Kolloquium, Heft 12 "Konstruktion", Ilmenau: T.H.I., 1967, p.69-74
109. Hubka, V., Theorie der Maschinensysteme, Berlin/Heidelberg: Springer, 1974
110. Hubka, V., "Konstruktionswissenschaft", VDI-Z **116,** (1974) Nr.11, p. 899-905, & Nr.13, p.1087-1094
111. Hubka, V., "Darstellen und Modellieren beim Konstruieren", Schw. Masch.**76** (1976), Nr.33, 35 & 37
112. Hubka, V., Theorie der Konstruktionsprozesse, Berlin/Heidelberg: Springer, 1976
113. Hubka, V., Konstruktionsunterricht an Technischen Hochschulen, Konstanz: Leuchtturm, 1978
114. Hubka, V.(ed), WDK 2 a & b: Bibliographie der Konstruktionswissenschaft, Zürich: Heurista, 1980 (an occasional serial publication, Bibliography of Design Science)
115. Hubka, V. et al., WDK 3: Fachbegriffe der wissenschaftlichen Konstruktionslehre in 6 Sprachen (Terminology of the science of design engineering in 6 languages), Zürich: Heurista, 1980
116. Hubka, V.(ed), WDK 5: Konstruktionsmethoden in Übersicht, Vol.1 of Proc. ICED 81 Rome, Milano: tecniche nuove, 1981
117. Hubka, V.(ed), WDK 6: Konstruktionsunterricht in Übersicht, Vol.2 of Proc. ICED 81 Rome, Milano: tecniche nuove, 1981
118. Hubka, V.(ed), WDK 8: Konstruktionsmethoden - Reading, Zürich: Heurista, 1983
119. Hubka, V., Principles of Engineering Design, London: Butterworth Scientific, 1982 (Translated and edited by Eder, W.E.,), reprint by Zürich: Heurista, 1987
120. Hubka, V.(ed), WDK 10: CAD, Design Methods, Konstruktionsmethoden, Proc. ICED 83 København, Zürich: Heurista, 1983 (2 vols)
121. Hubka, V., Andreasen, M.M., & Eder, W.E., WDK 4 b: Methodisches Konstruieren von Maschinensystemen - Fallbeispiele, Zürich: Heurista, 1983 (English edition in progress, see also [29, 122])
122. Hubka, V., & Andreasen, M.M., WDK 4 c: Methodisches Konstruieren von Maschinensystemen - Fallbeispiele, Zürich: Heurista, 1985 (English edition in progress, see also [29, 121])
123. Hubka, V., "Theorie Technischer Systeme - Eine Neue Disziplin für die Ingenieurausbildung", Int.Jnl.Appl.Eng.Educ., **1** (1985), p.153 158
124. Hubka, V.(ed), WDK 12: Theory and Practice of Engineering Design in International Comparison, Proc. ICED 85 Hamburg, Zürich: Heurista, 1985 (2 vols.)
125. Hubka, V., "A Curriculum Model Applying the Theory of Technical Systems", in [68], p.965-976
126. Hubka, V. & Schregenberger, J.W., "Paths Towards Design Science", in [68], p.3-14

127. Hubka, V., & Eder, W.E., "A Scientific Approach to Engineering Design", Design Studies **8** (1987), p.123-137
128. Jardine, A.K.S., Maintenance, Replacement and Reliability, London: Pitman, 1973
129. Johnson, R.C., Optimum Design of Mechanical Elements, New York: Wiley, 1961
130. Jones, J.Ch., Design Methods - seeds of human futures (2 ed), New York: Wiley, 1980
131. Kellermann, F., Mensch und Arbeit in der Industrie, Eindhoven: Phillips Techn. Bibl., 1964
132. Kesselring, F., Bewertung von Konstruktionen, Düsseldorf: VDI, 1951
133. Kesselring, F., Technische Kompositionslehre, Berlin/Heidelberg: Springer, 1954
134. Kirk, N.S., & Ridgeway, S, "Ergonomics Testing of Consumer Products", Appl. Erg., "Part 1: General Considerations" **1** No.5, (1970) p.295-300, "Part 2: Techniques" **2** No.1 (1971) p.12-18
135. Klaus, G., Wörterbuch der Kybernetik, Frankfurt a.M.: Fischer, 1969
136. Klir, G.J., An Approach to General Systems Theory, New York: Van Nostrand, 1969
137. Klir, G.J. (ed), Applied General Systems Research, New York: Plenum, 1978
138. Klir, G.J., Architecture of Systems Problem Solving, New York: Plenum, 1985
139. Klir, V., Cybernetic Modelling, London: Iliffe, 1967
140. Klose, J., "Theorie, Methodik und Unterstützung der Maschinenkonstruktion", Schw. Masch.**82** (1982), Nr.27
141. Koelze, H., "Alternativen für die Zukunft der Technik", VDI-Z **111** (1969), Nr.18
142. Koen, B.V., Definition of the Engineering Method, Washington, D.C.: A.S.E.E., 1985
143. Koller, R., Konstruktionsmethode für Maschinen-, Geräte- und Apparatebau (corrected reprint), Berlin/Heidelberg: Springer, 1979
144. Kourim, G., Wertanalyse, München: Oldenbourg, 1968
145. Krabs, W., Optimization and Approximation, Chichester: Wiley, 1979
146. Krick, E.V., An Introduction to Engineering and Engineering Design (2 ed), New York: Wiley, 1969
147. Kurajian, M.A., & Kurajian G.M., "Expanding Design Creativity", in [124], p.1075-1083
148. Layard, R.(ed), Cost-Benefit Analysis, Harmondsworth: Penguin, 1972
149. Leech, D.J., Management of Engineering Design, London: Wiley, 1972
150. Leyer, A., Machine Design, Glasgow: Blackie, 1974
151. Ljapunov, A.A., Systems Theory Research, New York: Consultants Bureau, 1973 (translated by E.S. Spiegelthal)
152. MacNiece, E.H., Industrial Specifications, New York: Wiley, 1953
153. Maslow, A.H., "A Theory of Human Motivation", Psychol. Rev.**50** (1943), p.370-396
154. Maslow, A.H., Motivation and Personality, New York: Harper & Row, 1954
155. Massey, B.S., Units, Dimensional Analysis and Physical Similarity, New York: Van Nostrand Reinhold, 1971
156. Matousek, R., Engineering Design, London: Blackie, 1963 (from German edition 1957)
157. Mayall, W.H., Industrial Design for Engineers, London: Iliffe, 1967
158. Miles, L.D., Techniques of Value Analysis and Engineering (2 ed), New York: McGraw-Hill, 1972
159. Mitroff, I., The Subjective Side of Science, Amsterdam: Elsevier, 1974 (Seaside, CA: Intersystems Publ., 1983)
160. Morrison, D., Engineering Design: the choice of favourable systems, London: McGraw-Hill, 1968
161. Mumma, A.G., "Technology's Motivating Force", Mechanical Engineering **89** Nr.4, (1967) p.18-21
162. Nadler, G., Work Systems Design: The Ideals Concept, Homewood: Irwin, 1967
163. Natrella, M.G., Experimental Statistics, NBS Handbook 91, National Bureau of Standards, Washington, DC: U.S. Govt. Print. Off., 1968
164. Naumann, F.K., Failure Analysis: Case Histories and Methodology, Metals Park, Ohio: Am. Soc. for Metals, 1983
165. Opitz, H., Werkstückbeschreibendes Klassifizierungssystem, Essen: Girardet, 1966
166. Osborne, A.F., Applied Imagination - Principles and Procedure of Creative Thinking, New York: Scribners, 1963
167. Pahl, G., & Beitz, W., Engineering Design, London: The Design Council, & Berlin/Heidelberg: Springer, 1984 (Translated and edited by K.M. Wallace)

168. Papanek, V., Design for the Real World, Toronto: Bantam 1973
169. Petroski, H., To Engineer is Human, New York: St. Martin Press, 1985
170. Pighini, U., "Methodical Design of Machine Elements", in [68], p. 254-264
171. Pinder, A., "Model of Models", Design Research (The Newsletter of the Design Research Society), Issue 16, April 1983
172. Planck, M. K. E. L., Der Kausalbegriff in der Physik (4 ed), Leipzig: J. A. Barth, 1948
173. Planck, M. K. E. L., Kausalgesetz und Willensfreiheit, Berlin/Heidelberg: Springer, 1923
174. Porter, A. W., The Method of Dimensions (2 ed), London: Methuen, 1943
175. Pronikov, A. S., Dependability and Durability of Engineering Products, London: Butterworths, 1973
176. Pugh, S., "Concept Selection - A Method that Works", in [116], p. 497-506
177. Rao, S. S., Optimization: Theory and Application (2 ed), New York: Wiley, 1984
178. Reuleaux, F., The Construktor, New York: Engineering Magazine, 1904 (translated by H. H. Suplee)
179. Reuleaux, F., & Moll, C. L., Constructionslehre für den Maschinenbau, Braunschweig: Vieweg, 1854
180. Riess, K., "Verfahrenstechnische Aufgaben des Ingenieurs bei der Gewinnung neuer und bei der Verbesserung vorhandener Produkte", VDI-Z **98** (1956), Nr. 23
181. Roadstrum, W., Excellence in Engineering, New York: Wiley, 1967
182. Rodenacker, W. G., Methodisches Konstruieren, Berlin/Heidelberg: Springer, 1970
183. Rolfe, S. T., & Barsom, J. M., Fracture and Fatigue Control in Structures, Englewood Cliffs, N. J.: Prentice-Hall, 1977
184. Ropohl, G., "Ansätze zu einer allgemeinen Systematik technischer Systeme", Schw. Masch. **76** (1976), Nr. 29
185. Ropohl, G., Eine Systemtheorie der Technik, München: Carl Hanser, 1979
186. Roth, K.-H., "Systematik der Maschinen und ihrer Elemente", Feinwerktechn. (1970), Nr. 11
187. Roth, K.-H., Konstruieren mit Konstruktionskatalogen, Berlin/Heidelberg: Springer, 1982
188. Rovida, E. (ed), WDK 14: Methodisches Konstruieren der Maschinenelemente, V. Hubka (gen ed), Zürich: Heurista, 1987
189. Schenck, H, Jr., Theories of Engineering Experimentation, New York: McGraw-Hill, 1961
190. Schlesinger, G., Testing Machine Tools for the Use of Machine Tool Makers, Users Inspectors & Plant Engineers (8 ed), Oxford: Pergamon, 1978
191. Schmidt, H., Lexikon der Philosophie, Stuttgart: Kröner, 1974
192. Shigley, J., Theory of Machines, New York: McGraw-Hill, 1965
193. Shigley, J. E. & Mitchell, L. D., Mechanical Engineering Design (4 ed), New York: McGraw-Hill, 1983
194. Simon, H. A., The Sciences of the Artificial, Cambridge, MA: MIT Press, 1969
195. Smith, C. O., Introduction to Reliability in Design, New York: McGraw Hill, 1976
196. Spunt, L., Optimum Structural Design, Englewood Cliffs, N. J.: Prentice Hall, 1971
197. Tellier, R. D., Operations Management, New York: Harper & Row, 1978
198. Thomson, V., & Graefe, U., "CIM - A Manufacturing Paradigm", National Research Council Canada, Div. of Mech. Eng. Report DM-6, NRC No. 26198, 1986/87
199. Tjalve, E., & Andreasen, M. M., "Zeichnen als Konstruktionswerkzeug", Konstruktion **27**, (1975) Nr. 1
200. Tjalve, E., Andreasen, M. M., and Schmidt, F. F., Engineering Graphic Modelling, London: Butterworths (1979)
201. Tjalve, E., A Short Course in Industrial Design, London: Newnes-Butterworths, 1979
202. Tomiyama, T. & Yoshikawa, H., "Extended General Design Theory", in H. Yoshikawa & E. A. Warman (eds), Design Theory for CAD, Proc. IFIP W. G. 5.2, Amsterdam: North Holland, 1987, p. 95-124
203. Towill, D. R., "Management Systems Applications of Learning Curves and Progress Functions", Eng. Costs and Prod. Econ. **9** (1985), p. 369-383
204. Truck, A., & Prüssmann, A., "Modell zur Beurteilung der Entwicklung neuer Produkte", VDI-Z **112** (1970), Nr. 2
205. Tschochner, H., Konstruieren und Gestalten, Essen: Girardet, 1954
206. Von Alven, W. H. (ed), Reliability Engineering (ARINC Research Corp.), Englewood Cliffs: Prentice-Hall, 1964

207. Wallace, K.M., & Hales, C., "Possible Future Trends in Engineering Design Research", in [124], p.899–909
208. Wearne, S.H., "A Review of Reports of Failures", Proc. I. Mech. E. **193,** (1979) p.125–136
209. Woodson, W.E., & Conover, D.W., Human Engineering Guide for equipment designers (2 ed), Berkeley, Calif: U. of C. Press, 1964
210. Wörgerbauer, H., Die Technik des Konstruierens, München: Oldenbourg, 1943
211. Vroom, V.H., & Deci, E.L. (eds), Management and Motivation, Harmondsworth: Penguin, 1975
212. Yoshikawa, H., "General Design Theory: Theory and Application", in Conference on CAD/CAM Technology in Mechanical Engineering, Cambridge, Mass.: M.I.T., 1981, p.370–6
213. Yoshikawa, H., "General Design Theory and CAD System", in T. Sata & E.A. Warman (eds) Man-Machine Communication in CAD/CAM, Proc. IFIP W.G.5.2/5.3 Working Conference, Amsterdam: North Holland, 1981, p.35
214. Zwicky, F., The Morphological Method of Analysis and Construction, Courant Anniv. Vol., New York: Wiley-Interscience, 1948

etrics of form 125
iniaturization 203
inimalization 206
ode 84, 148, 165
ode of action 19, 41, 59, 60, 61, 66, 72, 76, 77, 78, 81, 84, 91, 96, 109, 137, 213, 228, 256, 257
odel(-ling, -lable) 55, 56, 64, 76, 77, 101, 113, 138, 146, 147, 148, 161, 168, 172, 173, 177, 185, 216, 217, 218, 224, 230, 232, 248, 252, 259
odification 182
odular 84, 203, 204, 206
onitored 251
orphology (-ical) 74, 105, 219, 232
otion 71
otive (-ation) 28, 199, 208, 210, 228, 229, 232
otor 258

ture (-al) 7, 33, 36, 45, 59, 154, 157, 218, 249
ed (see also requirement) 19, 21, 23, 29, 35, 60, 70, 71, 87, 153, 163, 192, 199, 200, 201, 207, 214, 216, 228, 229, 250
wly designed 95
n-holonomic 133
rmative 147
velty 103, 152

EM 82
ject 6, 42, 80, 93, 215, 249, 252, 254, 258
jective(-ly) 110, 161, 163, 228, 257
server (-ing, -ation) 124, 218, 259
solete (-escence) 122, 189, 196
e-off 104, 111, 128, 183, 186
e-shot device 122
togenesis 214
erand 22, 26, 28, 29, 30, 33, 34, 35, 36, 37, 41, 42, 48, 49, 53, 57, 58, 60, 64, 65, 71, 90, 91, 93, 97, 105, 107, 109, 117, 121, 122, 141, 177, 183, 213, 215, 216, 228, 229, 246, 250, 251, 255, 257, 260
erate (-ing) 121, 148, 155, 183, 189
eration 18, 37, 43, 44, 57, 73, 74, 91, 149, 151, 173, 174, 178, 181, 184, 185, 186, 191, 203, 204, 216, 249, 250, 251, 258
erational 65, 81, 121, 137, 138, 141, 155, 163, 182, 192, 203, 205
erator 26, 29, 30, 34, 35, 36, 38, 44, 64, 107, 112, 123, 141, 177, 188, 213, 216, 229, 251
tical 115
timal (-izing) 101, 149, 161, 216, 218, 232
an 4, 60, 61, 66, 70, 73, 74, 78, 79, 87, 90, 92, 138, 204, 213, 256, 257, 259
an structure 61, 66, 70, 72, 75, 76, 77, 78, 79, 81, 85, 92, 137, 213, 256, 259
anism 61, 77, 78
anization 99
ginal(-ity) 95, 101, 103, 104, 191, 204
gin(-ation) 63, 91, 142, 174, 177, 178, 184, 185, 186, 191, 198, 249

output 26, 33, 34, 48, 57, 61, 65, 72, 78, 81, 90, 91, 107, 109, 122, 130, 155, 177, 222, 245, 246, 249, 251, 253, 255, 257, 258
out-sourced 82, 98, 103, 223
overhead 138

PERT 188, 231
paradigm solution (see master)
parallel 185, 188, 189, 253
parameter 42, 62, 103, 121, 129, 150, 153, 155, 192, 194, 247
partial 34, 37, 38, 46, 57, 61, 75, 92, 97, 118, 120, 132, 154, 165, 174, 183, 190, 191, 214, 223, 226, 230, 244, 246, 248, 259, 260
participating 90
parts list 66, 85, 177, 186
patent(-ability) 129, 152, 181, 182, 204, 209
path-time diagram 55
perceiving 148
performance 121, 160, 177, 192, 256, 259
perpetuum mobile 154
phases 261
phenomenon 33, 36, 45, 143, 217, 250, 252
philosophy 58
physical 49
physical elementary function 74
pilot plant 177
plan(-ning) 22, 30, 51, 147, 149, 183, 189, 208, 214, 232
plant 185
plastics 205
policy 84, 207
pollution 48, 65, 142
position 249
practice (see code of p.)
predicate 242, 247, 254
predicting 147, 149
preparing (-ation) 37, 39, 57, 99, 175, 177, 181, 183, 184, 261
prescribing (-ption, -ptive) 84, 147, 216, 220, 250
primitive 172, 173
principal flow (stream) 38
principle 5, 7, 27, 37, 74, 84, 92, 93, 96, 103, 112, 155, 176, 183, 194, 202, 213, 217, 221, 228
priority 154
probabilistic 59, 109, 249
problem(-solving) 113, 151, 176
procedure (-al) 161, 229, 258, 259
process 9, 13, 19, 22, 23, 26, 30, 33, 34, 35, 36, 37, 41, 44, 48, 53, 57, 60, 71, 89, 90, 91, 93, 95, 109, 141, 147, 151, 156, 164, 174, 177, 180, 188, 190, 191, 192, 204, 214, 215, 228, 229, 230, 249, 250, 251, 254, 257, 258, 261
processing 45, 93, 161, 165
product 22, 126, 153, 210, 232
production 99, 104, 136, 177, 182, 183, 184, 191
profitability 141
progression 254
project 95, 185

propelling 38, 40, 46, 57, 74, 76, 120, 203, 206, 222
property 26, 27, 36, 42, 49, 62, 72, 75, 76, 81, 82, 84, 85, 92, 95, 97, 101, 103, 124, 128, 144, 146, 147, 149, 152, 155, 156, 158, 160, 161, 163, 164, 165, 167, 168, 169, 173, 177, 181, 182, 183, 184, 185, 186, 189, 191, 192, 194, 195, 198, 202, 203, 205, 206, 210, 213, 215, 217, 228, 244, 245, 246, 247, 250, 252, 255, 256, 259, 260
prototype 104, 128, 147, 177, 181, 182, 183, 186, 256
psychological 49
purchased (-ing) 98, 103, 188, 223
purpose(-ful) 9, 23, 59, 60, 61, 65, 72, 74, 78, 83, 91, 93, 95, 107, 109, 110, 147, 149, 164, 168, 173, 194, 219, 221, 245, 254

quality (-ative) 42, 49, 102, 104, 107, 109, 110, 130, 135, 136, 144, 152, 163, 165, 167, 181, 186, 204, 205, 213, 247
quantify (-iable, -ication) 42, 108, 114, 147, 149, 164, 216, 218, 247, 248
quantity (see also measure) 42, 104, 109, 186

randomness 59
rating, ranking 53, 114, 165
rationalize (-ation) 102, 168, 183, 205
raw material 200
reaction 65
reading 149
realize (-ing) 34, 39, 60, 74, 75, 76, 92, 93, 117, 119, 146, 147, 149, 156, 157, 168, 176, 177, 181, 186, 192, 200, 204, 207, 210, 216, 228, 229, 245, 254, 256, 259
receptor 4, 76, 78, 90, 247, 257
recovery 142
re-cycle (-ing) (see also scrap, recovery, waste) 50, 142, 177, 206
re-design 95, 103, 157
reflexivity 252
regulating 38, 46, 57, 74, 76, 90, 117, 120, 123, 202, 203, 206, 207, 251, 259
rejection 165
relationship 9, 21, 23, 25, 30, 40, 44, 62, 66, 70, 72, 78, 82, 87, 93, 97, 109, 111, 112, 113, 115, 119, 132, 133, 143, 148, 149, 164, 169, 186, 191, 194, 216, 217, 219, 224, 226, 229, 244, 245, 246, 247, 251, 252, 253, 254, 255, 257, 259, 260
relative 109, 114, 157, 163
reliability 109, 121, 192, 203, 232
repair 185
replacement 122
represent(-ing, -ation) 19, 55, 71, 77, 81, 84, 113, 146, 168, 169, 173, 217, 226, 232, 250, 252
request for tender 151
requirement (see also need) 33, 49, 60, 70, 71, 81, 85, 87, 103, 110, 111, 129, 144, 146, 151, 153, 154, 155, 156, 158, 161, 164, 165, 168, 177, 181, 185, 186, 200, 204, 205, 210, 216, 223, 232, 249
requirements specification 111, 146, 151
research 39, 176, 180, 200, 201, 207, 208, 213
resolve 75
resources 155, 177
restriction 36
return 138
re-use 95, 96, 97, 103, 140, 142, 205, 206, 223
review (see design audit)
rheonomic 133
risk 104, 154, 161, 184, 200, 209
robotics 105, 130, 204
Ropohl 23
rule 250, 251
running re-design 182

SI 102
safety 20, 121, 184
salvage 142
sample 149
sample solution (see master)
saturation 195
scale (of measurement) 62, 108, 165, 247, 258
schematic (see also diagram) 76, 79, 180
science (-tific) 1, 10, 58, 59, 115, 138, 157, 199, 212, 2 229, 259
science of design (see design science)
scleronomic 133
scrap 142
secondary 48, 57, 65, 109, 122, 155
selecting 71, 164, 216
sequence (-tial) 18, 37, 63, 249
series 253
servicing 182
set 243
significance 164
similarity 93, 105, 147, 148, 252, 253
similitic 148
simulate 146, 149
size 131, 134
skill 99
society (-al, -etal) 1, 3, 23, 33, 49, 115, 154, 201, 202
socio-technical 23, 32, 35, 201, 207, 228, 230, 232
solid model 173
solution 71, 121, 229
solution searching for problem 209
space 39, 121
spatial relationship 78, 82, 224, 254
special theory 9, 56, 211, 212, 213, 221, 228
specialist (-ization) 98, 185
specification 36, 70, 71, 111, 151, 184, 229, 232
sponsor 152, 153
stable 250
standard(-ization) 96, 102, 103, 104, 129, 152, 182, 2 205, 206, 232, 251, 253, 259
standardized (-ing) 96, 102, 103, 146, 259
state 16, 22, 26, 29, 33, 35, 38, 42, 57, 59, 63, 77, 81, 110, 158, 168, 230, 245, 246, 248, 249, 251, 258, 260
state of embodiment 87, 90, 92, 110, 137, 158, 165, 2 247, 255, 260
state of the art 49, 50, 130, 152, 157, 192, 256

:atement 176, 219
:atic 249
atistics (see also probabilistic) 143, 232
ep(-wise) 42, 258
ochastic (see probabilistic)
oring 45, 93
rength 127, 157, 163
ructure 5, 9, 26, 42, 44, 57, 61, 62, 63, 66, 70, 71, 72, 75, 76, 77, 78, 79, 81, 85, 87, 91, 92, 97, 101, 103, 109, 111, 119, 131, 132, 137, 147, 169, 170, 176, 180, 181, 183, 185, 204, 213, 216, 245, 246, 249, 250, 251, 252, 255, 257, 260
yle (-ing) 124, 126
b-assembly 78, 79, 95, 120, 212
bjective 109, 110, 161, 163
bstantiate 146, 161, 219
pply 163
pport(-ing) 39, 46, 57, 65, 74, 76, 219, 246, 260
rface 131, 135, 172, 205
mbol(-ic) 147, 229, 242, 250
mmetry 252, 253
nthesis 76, 87, 229, 249
nthesizing characteristic 160, 161, 163, 164
stem 7, 8, 9, 21, 23, 27, 28, 30, 31, 35, 37, 48, 57, 58, 60, 64, 76, 90, 108, 117, 120, 134, 214, 219, 223, 224, 226, 230, 244, 245, 246, 249, 250, 258, 259, 260
stematic (see methodical)
stems engineering, theory 232, 242

chnik 1
k 60, 66, 81, 91, 111, 120, 123, 154, 190, 207, 213, 249, 254
onomy 105, 218, 223, 228
ching 228, 259, 260
hnical 6, 50, 51, 115, 156, 163, 199, 241
hnical level (see also state of the art) 176, 192, 195
hnical process (TP) 9, 30, 35, 36, 40, 48, 57, 60, 71, 0
nnical system (TS) 8, 9, 27, 28, 35, 37, 57, 58, 60, 64, 2, 77, 89, 90, 109, 128, 215
nnological principle 27, 34, 37
nnology 1, 13, 15, 21, 23, 26, 29, 34, 36, 44, 47, 50, 76, 4, 90, 99, 103, 118, 121, 175, 188, 213, 228, 229, 230, 41, 250, 251, 254, 260
nnology push, transfer 21, 93, 207
nnosphere 32
ology (-ical) 61, 221, 260
n(-inology) 5, 241, 242, 245, 247, 251, 254
(-ing) 104, 182, 184, 186, 218, 232
ry (-ize) 4, 56, 148, 212, 213, 218, 254

theory of machines 5
theory of technical systems 8
thermal 115
time 26, 39, 42, 186, 254
title 96
tolerance 131, 132, 136, 169, 205, 232
tool 47
topology 226, 228, 254
total function 75, 79
tort 129
training 229
transfer function 72
transform(-ation) 18, 22, 23, 28, 29, 33, 34, 45, 46, 53, 58, 65, 90, 91, 93, 111, 117, 121, 122, 131, 138, 176, 177, 183, 215, 216, 230, 250, 251, 261
transformation process 35, 44, 53, 57, 109, 214, 251
transformation system 23, 32, 34, 58, 64, 76, 208, 213
transformation within TS 29
transitivity 252
transmitter 4
transport(-ing) 45, 93, 182
trial 186
trigger 59
type 38, 104, 248
typifying (-ication) 102, 205, 206

union 244
unit 109, 110, 164, 247, 258
unit operation 44
usage 137, 159, 163, 195, 197
user 129

value 21, 50, 107, 108, 156, 157, 160, 161, 163, 165, 167, 177, 181, 191, 192, 203, 206, 247, 248, 251, 255, 258, 259, 260
value analysis 40, 50, 232
variable 113
variant (see alternative)
verbal 55, 147, 229
verify(-ing, -ication) 39, 104, 136, 147, 161, 181, 216, 231, 259

WDK vi, 254
waste 65, 122, 142, 247
wear(-out) 121, 184, 189
weighted (-ing) 114, 165
wire-frame model 172
working 44, 103, 119, 141, 155, 176, 177, 181, 183, 185, 186, 188, 191, 199, 202, 203, 213, 215, 251
working means (designer's) 73, 84, 200, 216, 220

V. Hubka

Principles of Engineering Design

WDK 1: W. E. Eder (Ed.)

Translated from the German by W. E. Eder

2nd edition. 1988. 26 figures. Approx. 130 pages.
ISBN 3-540-50126-6
Originally published by Butterworth & Co., 1982.

From the preface:
Methodical procedures in design engineering, in spite of their relatively recent use, have taken on a wide variety of forms. Is this multiplicity only an external characteristic, or is it an integral part of the problem? This question represents a challenge, and has stimulated a search for the common denominator in the efforts to date.
One result of this search – the General Procedural Model of Design Engineering – can be viewed as an attempt to synthesise published opinions, contributions to discussions by students and designers, and experiences from teaching in colleges and in continuing education courses.
In preparing this work, attention was given to the following requirements:

- as far as possible, general applicability to the area of machine systems;
- as far as possible, consistent treatment of each step, and consistent terminology;
- a rational foundation for all steps and for its sequence;
- presentation of the fundamental knowledge that can assist understanding of the procedures;
- minimum of descriptive content, to ensure adequate clarity, and to preserve widest generality.

Based on these requirements, I hope that the model will assist both the student and the practising design engineer.

Springer-Verlag
Berlin Heidelberg New York
London Paris Tokyo

Springer